FACILITATING CLIMATE CHANGE RESPONSES

A REPORT OF TWO WORKSHOPS ON KNOWLEDGE FROM THE SOCIAL AND BEHAVIORAL SCIENCES

Paul C. Stern and Roger E. Kasp

Panel on Addressing the Challenges of Climate Change
Through the Behavioral and Social Sciences

Committee on the Human Dimensions of Global Change

Division of Behavioral and Social Sciences and Education

NATIONAL RESEARCH COUNCIL
OF THE NATIONAL ACADEMIES

THE NATIONAL ACADEMIES PRESS
Washington, D.C.
www.nap.edu

THE NATIONAL ACADEMIES PRESS 500 Fifth Street, N.W. Washington, DC 20001

NOTICE: The project that is the subject of this report was approved by the Governing Board of the National Research Council, whose members are drawn from the councils of the National Academy of Sciences, the National Academy of Engineering, and the Institute of Medicine. The members of the committee responsible for the report were chosen for their special competences and with regard for appropriate balance.

This project was supported by Grant No. 2008-2146. Support of the work of the Committee on the Human Dimensions of Global Change is provided by the William and Flora Hewlett Foundation. Any opinions, findings, conclusions, or recommendations expressed in this publication are those of the author(s) and do not necessarily reflect the views of the sponsor.

International Standard Book Number-13: 978-0-309-16032-2
International Standard Book Number-10: 0-309-16032-4

Additional copies of this report are available from the National Academies Press, 500 Fifth Street, N.W., Lockbox 285, Washington, DC 20055; (800) 624-6242 or (202) 334-3313 (in the Washington metropolitan area); Internet http://www.nap.edu.

Printed in the United States of America

Suggested citation: National Research Council. (2010). *Facilitating Climate Change Responses: A Report of Two Workshops on Knowledge from the Social and Behavioral Sciences.* P.C. Stern and R.E. Kasperson, Eds. Panel on Addressing the Challenges of Climate Change Through the Behavioral and Social Sciences. Committee on the Human Dimensions of Global Change, Division of Behavioral and Social Sciences and Education. Washington, DC: The National Academies Press.

THE NATIONAL ACADEMIES

Advisers to the Nation on Science, Engineering, and Medicine

The **National Academy of Sciences** is a private, nonprofit, self-perpetuating society of distinguished scholars engaged in scientific and engineering research, dedicated to the furtherance of science and technology and to their use for the general welfare. Upon the authority of the charter granted to it by the Congress in 1863, the Academy has a mandate that requires it to advise the federal government on scientific and technical matters. Dr. Ralph J. Cicerone is president of the National Academy of Sciences.

The **National Academy of Engineering** was established in 1964, under the charter of the National Academy of Sciences, as a parallel organization of outstanding engineers. It is autonomous in its administration and in the selection of its members, sharing with the National Academy of Sciences the responsibility for advising the federal government. The National Academy of Engineering also sponsors engineering programs aimed at meeting national needs, encourages education and research, and recognizes the superior achievements of engineers. Dr. Charles. M. Vest is president of the National Academy of Engineering.

The **Institute of Medicine** was established in 1970 by the National Academy of Sciences to secure the services of eminent members of appropriate professions in the examination of policy matters pertaining to the health of the public. The Institute acts under the responsibility given to the National Academy of Sciences by its congressional charter to be an adviser to the federal government and, upon its own initiative, to identify issues of medical care, research, and education. Dr. Harvey V. Fineberg is president of the Institute of Medicine.

The **National Research Council** was organized by the National Academy of Sciences in 1916 to associate the broad community of science and technology with the Academy's purposes of furthering knowledge and advising the federal government. Functioning in accordance with general policies determined by the Academy, the Council has become the principal operating agency of both the National Academy of Sciences and the National Academy of Engineering in providing services to the government, the public, and the scientific and engineering communities. The Council is administered jointly by both Academies and the Institute of Medicine. Dr. Ralph J. Cicerone and Dr. Charles M. Vest are chair and vice chair, respectively, of the National Research Council.

www.national-academies.org

PANEL ON ADDRESSING THE CHALLENGES OF CLIMATE CHANGE THROUGH THE BEHAVIORAL AND SOCIAL SCIENCES

ROGER E. KASPERSON (*Chair*), George Perkins Marsh Institute, Clark University
RICHARD N. ANDREWS, Department of Public Policy, University of North Carolina
MICHELE M. BETSILL, Department of Political Science, Colorado State University
STEWART J. COHEN, Adaptation and Impacts Research Section, Environment Canada, and Department of Forest Resources Management, University of British Columbia
THOMAS DIETZ, Environmental Science and Policy Program, Michigan State University
ANDREW J. HOFFMAN, Ross School of Business, University of Michigan
ANTHONY LEISEROWITZ, School of Forestry and Environmental Studies, Yale University
LOREN LUTZENHISER, Department of Urban Studies and Planning, Portland State University
SUSANNE C. MOSER, Susanne Moser Research and Consulting, Santa Cruz, California
GARY W. YOHE, Department of Economics, Wesleyan University

PAUL C. STERN, *Study Director*
LINDA DePUGH, *Administrative Assistant*

COMMITTEE ON THE HUMAN DIMENSIONS OF GLOBAL CHANGE
2010

Preface

A series of reports from the National Research Council (NRC), including the recent reports in the "America's Climate Choices" project, has underscored what has long been known—that the U.S. government continues to lack the basic social science capability to address many of the nation's serious environmental and health problems. It is increasingly clear that the technological systems facing the United States—in such varied domains as energy systems, chemicals in the environment, radioactive waste disposal, nanotechnology, and health systems—are at root social-technical systems whose prospects and problems need to be addressed with integrated analysis of technology, people, ecology, and social institutions. Yet the interactions of technology and environment with human systems remain a neglected research area.

The recent NRC review of the U.S. Climate Change Science Program, Restructuring Federal Climate Research to Meet the Challenges of Climate Change, is one of a plethora of examples. The program across 13 federal agencies lists social science issues as central to two to three of its major goals. Yet the review found that although effective progress had been made on the natural science issues, very little progress had been achieved on such critical social and behavioral science issues as risk communication, potential impacts of climate change on human systems, and engaging stakeholders. Ant it was apparent that social science issues commanded only a few percent of the total budget for federal climate science budget. In another case, the National Oceanic and Atmospheric Administration (NOAA) recently completed a review of its progress since a highly negative report in

2004 on its social science capability detailed its inadequate expertise and resources. The 2009 review, Integrating Social Science into NOAA Planning, Evaluation and Decision Making: A Review of Implementation to Date and Recommendations for Improving Effectiveness, found that not only had NOAA failed to make significant progress, it had actually lost ground over the 5-year period.

To help show federal agencies the value of the social and behavioral sciences, the Hewlett Foundation provided a modest grant to the NRC for two workshops that would showcase how the behavioral and social sciences could contribute useful knowledge to the nation's effort to respond to climate change. The first workshop, in December 2009, was on mitigation of climate change risks, and the second, in April 2010, was on adaptation to climate change.

The terms "mitigation" and "adaptation" have particular uses in the climate change science field that are different from usage common in the social sciences. In climate change science, mitigation usually refers to activities that reduce the extent of climate change, and adaptation refers to actions that reduce the damage resulting from whatever climate change occurs. In hazard research in the social sciences, by contrast, mitigation typically refers to the amelioration of consequences by whatever means. Adaptation, in climate change science, usually refers to organized efforts rather than what might be termed "autonomous adaptation" by individual natural resource managers. Indeed, as several speakers at the adaptation workshop made clear, adaptation is ongoing in all societies where change in their social and natural environments are an inherent part of daily decision making and economic behavior.

The term "social-technical" system seems particularly relevant here. The workshops gave particular attention to the couplings or linkages that shape the interactions among society, technology, and ecology. At the same time, it is apparent that the understanding of the nature of this coupling, and particularly its causal connections, is still highly rudimentary. Much work is needed for a more robust understanding of how such coupling comes into being and how it functions in different contexts.

Social science research on mitigation and that on adaptation are at different stages of development. Work on mitigation includes several bodies of knowledge that can offer useful practical insights now. For example, research on household behavior revealed significant opportunities for greenhouse gas reductions without major changes in values or lifestyles. For adaptation, however, although much is known about specific elements or facets of the processes of adaptation, no general integrative theory or framework now exists to guide a coherent research agenda or to suggest a set of best practices. The workshops were a salutary experience for most of the participating social and behavioral scientists, often because there was the opportunity to dig deeply into issues with other accomplished researchers working on global climate change. The workshops also benefited strongly from the attendance, and commentary, of a number of agency representatives, including those from the National Oceanic and Atmospheric Administration; National Science Foundation; Office of Science and Technology Policy; U.S. Agency for International Development; U.S. Departments of Agriculture, Education, Energy, Interior, and Transportation; U.S. Environmental Protection Agency; U.S. Government Accountability Office; and U.S. Global Change Research Program.

I am very grateful to the staff of the Committee on the Human Dimensions of Global Change: Paul Stern, director of the committee and of this project, and Linda DePugh, who handles all the administrative arrangements. In addition, the project also would like to thank NRC fellow Hadas Kushnir who provided research support for the second workshop. For background, the project commissioned a paper from Seth Tuler of the Social and Environmental Research Institute, which overviewed a number of the major studies conducted on the siting of controversial facilities. The paper, "Factors Influencing Public Support and Opposition to Renewable Energy Facility Siting: A Review of the Literature," will inform future efforts by the Committee on the Human Dimensions of Global Change to develop new projects that build on the results of these workshops.

This report has been reviewed in draft form by individuals chosen for their diverse perspectives and technical expertise, in accordance with procedures approved by the NRC's Report Review Committee. The purpose of this independent review is to provide candid and critical comments that will assist the institution in making its published report as sound as possible and to ensure that the report meets institutional standards for objectivity, evidence, and responsiveness to the study charge. The review comments and draft manuscript remain confidential to protect the integrity of the deliberative process.

We thank the following individuals for their review of this report: Arun Agrawal, School of Natural Resources and Environment, University of Michigan; Kristie L. Ebi, Department of Global Ecology, Carnegie Institution for Science, Stanford, CA; and Charles Wilson, Department of Geography and Environment, London School of Economics and Political Science, University of London. Although the reviewers provided many constructive comments and suggestions, they were not asked to endorse the report nor did they see the final draft of the report before its release. The review of this report was overseen by Susan Hanson, Department of Geography, Clark University. Appointed by the NRC, she was responsible for making certain that an independent examination of this report was carried out in accordance with institutional procedures and that all review comments were carefully considered. Responsibility for the final content of this report rests entirely with the authoring panel and the institution.

Roger E. Kasperson, *Chair*
Panel on Addressing Change Through the
Behavioral Social Sciences

Contents

Appendixes

Introduction

Scientists and policy makers increasingly recognize global warming and other aspects of climate change as significant threats to the future of Earth's ecosystems and to human well-being. If left unchecked, climate change could lead to worsening consequences, including faster rising sea levels; more floods, storms, fires, and waterborne and vector-borne diseases; heat-related illness; crop failures; shifting ecosystems; and environmental degradation. Although scientists still disagree in their estimates of the timing and magnitude of particular consequences, there is widespread agreement that the risks are sufficiently serious to warrant action to reduce the net future human influence on climate (mitigation) and to promote successful adaptation to the consequences of climate change that cannot be avoided (National Research Council, 2010a, 2010b, 2010c; U.S. Global Change Research Program, 2009b).

Responding to climate change requires an expansion of the range of scientific work on climate change. This is a "new era of climate change research" (National Research Council, 2010b:4), one that requires a much stronger emphasis than previously on the understanding of human-environment systems and a much greater integration of the social and behavioral sciences with the other sciences concerned with climate change. Much of the expanded research agenda is directed to use-inspired fundamental research (Stokes, 1997) that can support effective human responses to climate change, including efforts to limit its magnitude and to adapt to its consequences.

The William and Flora Hewlett Foundation, understanding the need for these kinds of research and the need for policy makers at the national level to entrain the behavioral and social sciences in addressing the challenges of global climate change, called on the National Research Council (NRC) to organize two workshops in Washington, DC, to showcase some of the decision-relevant contributions that these sciences have already made and can advance with future efforts. The Panel on Addressing the Challenges of Climate Change Through the Behavioral and Social Sciences was formed to organize the workshops under the auspices of the NRC's Committee on the Human Dimensions of Global Change. The workshops were held on December 3-4, 2009, and April 8-9, 2010.

The panel was asked to organize workshops in two broad areas in which insufficient attention has been paid to the potential contributions of behavioral and social sciences: (1) mitigation (behavioral elements of a strategy to reduce the net future human influence on climate) and (2) adaptation (behavioral and social determinants of societal capacity to minimize the damage from climate changes that are not avoided). The workshops were intended to demonstrate the contributions that the behavioral and social sciences can make for more effective responses to climate change. It was also intended that the workshops would lay the foundation for further inquiries.

The panel developed and considered a number of topical areas for discussion before settling on the agendas, topics, and invited presenters for the two workshops. There are fairly large and well-developed social and behavioral science literatures on several aspects of climate change mitigation, and not all of them could be covered in a two-day workshop. We decided to focus on a few issues we thought would be particularly relevant to current policy debates. One of the issues is public understanding of climate change—a topic that is important both for mitigation and adaptation. We believed that a scientific examination of how nonscientists think about climate change could help explain shifts in public opinion and levels of public support for climate policies and could be useful for improving public understanding and for educating the next generation of citizens on the topic. We devoted a half-day session to this topic.

We organized additional half-day sessions around three other topics: (1) the potential for mitigating climate change through household action, (2) public acceptance of energy technologies, and organizational change, and (3) the "greening" of business. In each session, presenters reported on the knowledge base on the topic, and invited discussants and other participants considered the implications of the findings for policy choices.

The workshop on adaptation to climate change took a different form because of the different state of social and behavioral science knowledge. Multidisciplinary research on adaptation to natural climate variations has been conducted for decades at a relatively low level of intensity. However, the issue of adaptation to anthropogenic climate change has only relatively recently become a major one on research and policy agendas. Thus, the research literature and agenda are less well defined and more dispersed across several disciplines and related fields than those on mitigation.

To plan the adaptation workshop, we began by contacting a number of policy makers in federal agencies who have been working with decision makers at federal, state, and local levels who are confronted with the need to take climate change into account in their work. We asked them what they would like to learn from research on adaptation and, with that input, we developed a list of key questions about climate change adaptation to pose to social and behavioral scientists. We invited researchers who had studied topics that we believed could shed light on these questions and who had directly examined multiple cases of adaptation. We asked them to report on what they had learned, and panel members volunteered to listen to these presentations and report at the end of the workshop on what they had heard during the workshop that might answer the decision makers' questions.

The workshops brought together leading researchers from across the behavioral and social sciences whose expertise and research can help address timely questions about responding to climate change. The presentations emphasized current research, some of it not yet published at the time it was presented. We found the discussions enlightening and stimulating, and we believe that, even in this written summary form, they will be useful to readers who are interested in the latest knowledge about human responses to climate change. The workshop material concretely illustrates some of the ways the behavioral and social sciences can contribute to the new era of climate research called for in the report *Advancing the Science of Climate Change* (National Research Council, 2010b). It also shows how these sciences can help in addressing the challenges of climate change.

This report does not present any conclusions, lessons, or the like as consensus statements by the panel. Although individual members of the panel drew conclusions from the workshops, some of which are mentioned in the report, it was not the purpose of these workshops to draw overall conclusions. Readers will have to do that for themselves.

It is important to be specific about the nature of this report, which documents the information presented in the workshop presentations and discussions. Its purpose is to lay out the key ideas that emerged from the workshops and should be viewed as an initial step in examining the research and applying it in specific policy circumstances. The report is confined to the material presented by the workshop speakers and participants. Neither of the workshops nor this summary is intended as a comprehensive review of what is known, although each generally reflects of the literature. The presentations and discussions were limited by the time available for the workshops.

Although this report was prepared by the panel, it does not represent a consensus of the panel. Rather, the report summarizes views expressed by workshop participants, and the panel is responsible only for its overall quality and accuracy as a record of what transpired at the workshops.

PLAN OF THE REPORT

The structure of this report reflects the organization of the two workshops. Part I summarizes the December 2009 workshop on public understanding and mitigation of climate change. Part II summarizes the April 2010 workshop on adaptation to climate change. Appendix A presents the agenda and list of participants of the December 2009 workshop, and Appendix B does the same for the April 2010 workshop. Appendix C presents biographical sketches of the panel members and staff.

Part I

Public Understanding and Mitigation of Climate Change

The December 2009 workshop was devoted to four distinct topical sessions and might therefore be considered as a set of smaller workshops. The first was devoted to public understanding of climate change and the other three to policy-related topics concerning efforts to limit future climate change by reducing greenhouse gas emissions or implementing low-emission technologies.

Roger Kasperson, the panel chair, introduced the December 2009 workshop by saying that this set of workshops is atypical of a National Research Council event in two ways. First, it allows a core of social scientists to engage in detailed discussion of the social science issues. Second, instead of formulating research agendas, its focus is on a few areas in which researchers are confident that the social sciences already know quite a bit that can contribute to policy discussions internationally, in the federal government, in the private sector, and at the state and local levels, both now and in the future. Chapters 1-4 report on the presentations and discussions at the December workshop:

1. public understanding of climate change,
2. opportunities for climate change mitigation by household action,
3. public acceptance of energy technologies, and
4. organizational change and the greening of business.

1

Public Understanding of Climate Change

Decades of research on climate have made it increasingly clear to Earth scientists that Earth's "climate is changing, and that these changes are in large part caused by human activities" (National Research Council, 2010b:1). However, these conclusions have recently lost support among the U.S. public. Similarly, scientists find that "climate change . . . poses serious risks for both human societies and natural systems" and that actions to mitigate (slow down) and adapt to these changes are needed and urgent (National Research Council, 2010b:1). This understanding and level of concern are not yet evident in national public policy. These divergences between science and society raise important social science questions: answers to those questions can help the nation make progress in dealing with climate change.

This chapter summarizes research that helps to answer some of these questions by explaining why understanding and responding to climate change have been so difficult. Anthony Leiserowitz's research describes "six Americas," each characterized by a unique set of understandings of and responses to climate change, some sharply at variance with climate science. Susanne Moser puts these differences in perception, understanding, and behavior in a broader context of societal forces. She discusses multiple reasons why climate change is hard for nonspecialists to understand, including that the topic is inherently difficult and complex; that understanding it requires a kind of cognitive functioning that does not come easily to most people; and that the media and society have not sent clear signals for the need to respond for the common good. The studies by Daniel Read and his colleagues find that many people's mental models of climate change are inadequate or different from scientific understanding and that people's confusions have persisted for almost two decades despite education and communication on the issue. Elke Weber discusses how ingrained cognitive and affective responses to risk can lead people astray when they consider the risks of climate change. Finally, the research reported by Riley Dunlap shows how an organized climate change denial "counter-movement" linked to conservative political institutions and elements of the fossil fuel industry has worked to influence public understanding and how an increasing ideological polarization in U.S. public opinion on the topic has followed their efforts. The chapter concludes by summarizing a discussion that considered how ongoing structural changes in the mass media might affect the potential to improve public understanding and what might be done to improve understanding through the education system and in the broader society.

INTRODUCTORY COMMENTS

Anthony Leiserowitz[1]
Yale University

Anthony Leiserowitz began by noting that the topic is of great current interest. Public support or its lack is clearly a constraint on national climate legislation, and global public opinion will affect action at the December 2010 Copenhagen meeting. Polls in late 2009 suggested that Americans were somewhat less convinced than before that climate change is real and human caused. The recent scandal of hacked e-mails, centered on the University of East Anglia, may also have influenced public opinion. Leiserowitz offered three comments to frame the discussion.

First, decades of research by natural scientists have tremendously improved understanding of how the climate system works, showing unequivocally that Earth is warming, that human activities are the primary cause, and that impacts are already beginning to be felt, with stronger ones expected in the future. Now, however, climate change is a major problem for the social and behavioral sciences, because it is rooted in the factors that drive human decision making and behavior and because the solutions to climate change will require human beings to choose and act differently as individuals, families, communities, nations, and societies. In addition, many of the impacts of greatest concern are the potential consequences of climate change for human well-being.

Second, the greatest source of uncertainty in climate models is future

[1]Presentation is available at http://www7.nationalacademies.org/hdgc/Session_Moderator_Public%20Understanding_of_Climate_Change_Anthony_Leiserowitz.pdf [accessed September 2010].

human behavior. Whether the world stabilizes global warming at 2 degrees Celsius or warming reaches a much higher level depends fundamentally on whether humans act to alter the trajectory of climate change. Thus, the social sciences are key to meeting the climate challenge. Furthermore, human systems are much more complex and hard to predict than the climate system. Carbon dioxide molecules all behave in the same way and do not change their behavior appreciably when scientists study them, but human beings do. Human beings as individuals and as societies are capable of a wide range of potential responses, making it very difficult to predict how they will respond to future climate change or to research on climate change.

Third, the U.S. public is not homogeneous. Leiserowitz and his colleagues have identified six distinct groups or segments (what they have termed "The Six Americas"), each of which responds in a very different way to climate change (Leiserowitz et al., 2008). They have labeled the segments as "alarmed" (18 percent of the population), "concerned" (33 percent), "cautious" (19 percent), "disengaged" (12 percent), "doubtful" (11 percent), and "dismissive" (7 percent). These groups vary in terms of how much they believe global warming is a reality, how concerned they are, and how motivated they are to take action. He emphasized that public response to climate change is not a linear response to scientific information. Rather, people are already predisposed either to accept or reject what scientists say about it and, similarly, to support or oppose proposed policies.

THE TROUBLE WITH CLIMATE CHANGE: WHY IS CLIMATE CHANGE HARD TO UNDERSTAND?

Susanne Moser[2]
Susanne Moser Research and Consulting

Susanne Moser began by stating that climate change as a phenomenon has attributes that make it extremely difficult for nonspecialists to understand. She posed a series of questions. Is the difficulty of understanding climate change in the nature of the topic? Is there a problem with how human brains are "wired"? Does the ultimate challenge lie in people's world views, which don't allow them to see climate change as a problem or to accept proposed solutions to it? Is the problem a failure of communication? Does the problem lie with how the mass media work and describe the issue? Are there too many more pressing distractions that preoccupy people's atten-

[2]Presentation is available at http://www7.nationalacademies.org/hdgc/Why%20is%20Climate%20Change%20Hard%20to%20Understand%20Susanne%20Moser%20SM%20Consulting.pdf [accessed September 2010].

tion? Finally, does it actually matter whether the public understands? She concluded that there are partial answers to all of these questions, and that the challenge of engaging the public on climate change involves a "perfect storm" of all of these factors operating together.

The difficulty of understanding is partly due to the characteristics of the problem. First, you cannot see climate change. You cannot see carbon dioxide: if it made the sky black, it might be more easily noticed. Second, change is happening very slowly on the time scale of human perception. People can easily remember one cold winter, but they cannot notice a sea level rise at the rate of 1 mm/year. And a driver from the city into the suburbs experiences a temperature change due to leaving the urban heat island that is larger than has been seen for the Earth over the past century. In addition, there is the perception that many of the impacts are distant in time and space. Public opinion polls in the United States and some other countries show that respondents see greater harm from global warming coming to animals, plants, and people and things that are far away from them than to those close to them (see Figure 1-1). Next, modern humans are insulated from their environments. People spend 20 hours or more per day in buildings. A time survey conducted in 2000 indicated that 51 percent of Americans spend no time outside, and an additional 30 percent spend no more than one hour outdoors. All of this makes people highly dependent on mediated communication for information on climate change. In addition, there is no quick, simple, or good fix. Taking action gives no gratification or very highly delayed gratification. A recent study by Solomon et al. (2009) shows that, short of active interference with the climate system to take carbon back out of the atmosphere, the climate will not return to the preindustrial state in the lifetime of anyone now living.

Climate change is also challenging for the human brain. People tend to react quickly to things that stem from the ill intent of an identifiable actor, that provoke moral outrage, that present clear and present danger, and that happen fast. Moser quoted Harvard psychologist Daniel Gilbert as saying that the ability to duck what is not yet coming is a stunning but recent innovation. Also, our impact outstrips our brain. Many people believe and everyone hopes that climate change is just the result of a natural cycle. Also, it is difficult to understand systems, which must be understood in order to comprehend the nature and magnitude of change that is needed to limit climate change. In addition, most people are much better at intuitive information processing than systematic processing. They have trouble dealing with uncertainty, so that uncertainty about climate, for many, provides a rationale for postponing action. It is demanding to deal not only with the overwhelming scientific complexity, but also with the moral complexity of the climate issue. Furthermore, the signals indicating that responses are needed are inadequate.

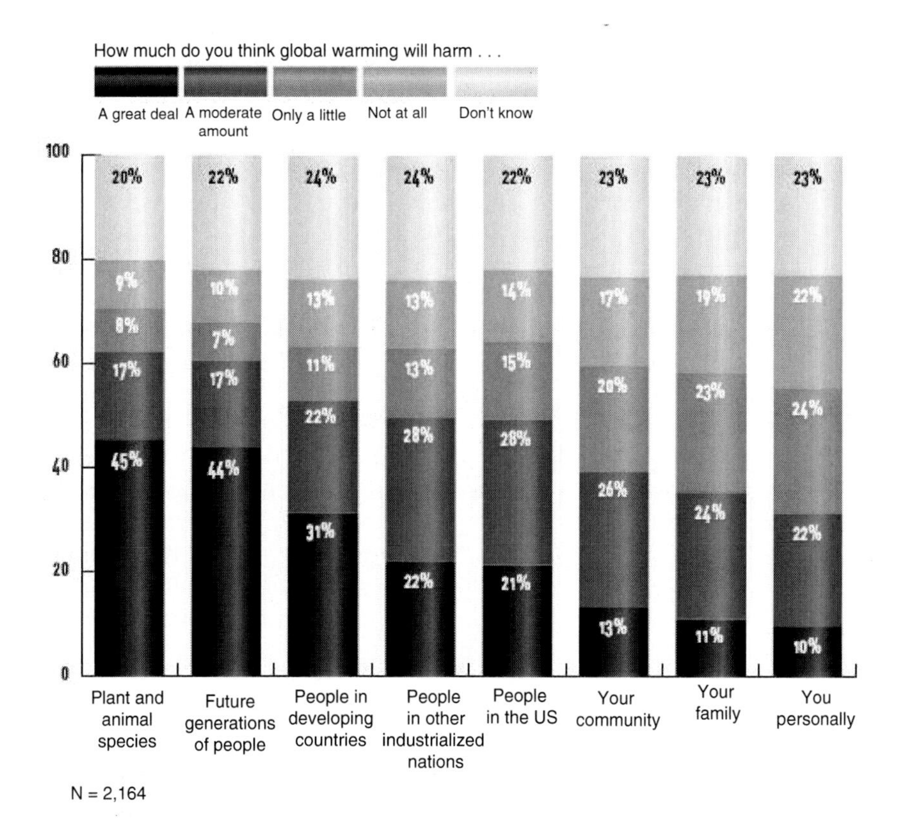

FIGURE 1-1 Perceived degree of harm from global warming to various entities, perceived by U.S. respondents.
SOURCE: Leiserowitz et al. (2009:30). Reprinted with permission.

Moser said that society has failed to signal the need for change. Climate change is the ultimate market failure. We have no price on carbon to signal the value of reducing emissions, no uniform and steady messaging, a lack of consistency between what leaders say and do, and a lack of social narratives that portray "climate protection" as a source of a socially desirable identity. Such signaling would be necessary to help people see beyond narrow self-interest and act for the common good. People filter information through a "cultural" lens colored by their general beliefs about society and about right and wrong. This filter operates prior to facts and shapes the interpretation of information. Moser remarked that she sees this filtering as the underlying cause of the "six Americas" reported by Leiserowitz. Particularly, the people at the extremes of that continuum of views say they are highly unlikely to change their views, probably because of their strong attachments to preexisting cultural world views.

Moser drew several implications from this analysis. People reject information that contradicts their beliefs and selectively attend to information that confirms them. This process leads to social and political polarization. Moser suggested that people need different forums for deliberation so they can understand each others' world views and seek common ground.

She noted that climate change is presented within multiple "frames," each of which works for some audiences, but none of which works for all. Some frames (such as some scientific ones) fail to resonate with audiences, yet when climate change is framed as a catastrophe or a threat to the social system, it may threaten some people's sense of self.

The media also are part of the problem. The traditional media are organized for profit, not for education. This fact shapes their choice of stories. Techno-cultural and economic changes in the media industry are part of the picture. They include a change from broadcasting to "narrow-casting" focused on selected audiences. The downsizing of the reporter corps, including on environmental and science issues, results in a greater probability of reporting with factual mistakes and shallower treatment or simply less climate change coverage. The mass media—as corporate businesses—focus on news that "sells," which puts a premium on extreme events, human interest stories, and controversies over slowly developing stories. The rise of the Internet and "social media" democratizes, but people are not exposed to ideas across the spectrum and tend to get news more frequently in sound bites and through peripheral information processing or from sources that conform to their preexisting views. Mass media can do less than they once could, and they are not good at direct persuasion, fostering behavior change, promoting two-way communication, dealing with issues in depth, or resolving conflicts, although some of these are much needed in climate change policy.

Finally, Moser offered a list of what climate change is not, as a way to explain why it is hard to keep the topic on the public agenda. To many people, climate change is not visible and so may not seem real. It is not immediate, even in its threat, cost, or the pleasure or satisfaction of trying to solve the problem. It is not (yet) relevant in the sense of being personal, here and now. It is not intuitively understandable. It is not easy to talk about at the dinner table. It is not easily solvable, which would provide a sense of personal and response efficacy. It is not morally simple. And it is not yet seen or perceived as a threat to everything people are and value.

In the discussion that followed the presentation, Richard Andrews noted that climate response is being given positive frames in some states, for example, as a development issue. Moser responded that the climate change discussion has been framed mainly by scientists and environmentalists, even though other framings are valid for certain audiences.

A questioner asked whether audiences can be segmented to focus differently on groups that have different views on the topic. Moser replied that an economic framing cuts across the population and that, for some audiences, an environmental justice frame can engage people with climate change. However, she doubted that there is a single frame that would work for everybody. Leiserowitz commented that an energy frame works for almost everyone because there is widespread concern about the nation's energy future.

NOW WHAT DO PEOPLE KNOW ABOUT CLIMATE CHANGE? A STUDY OF LAY BELIEFS AND MENTAL MODELS

Daniel Read[3]
Yale University

Daniel Read presented work he did in collaboration with Ann Bostrom, Rebecca Hudson, Anu Narayanan, and Travis Reynolds. An earlier related project in 1992 in which he, Bostrom, and others participated resulted in two papers (Bostrom et al., 1994; Read et al., 1994). Since then, popular sources have bombarded people with information and arguments on climate change. In a new study, the research group replicated the methods from 1992 using very similar people under similar circumstances in 2009 (in both cases, July 4, in the same park in Pittsburgh). (Note: This study, Reynolds et al., 2010, was accepted for publication after the workshop.) Its goal was to see whether and how public understanding had changed. The 2009 sample scored a bit lower on education level and was augmented by a more highly educated subsample for comparability.

The results showed little change from 1992 in beliefs about whether human actions have changed global climate—if anything, there was a decline in this belief. Understanding of the nature of the greenhouse effect has increased only slightly in 17 years. As in 1992, people were asked how much temperature change there has been to date, how much they would expect in 10 years, and how much they would expect by 2050. The results were highly similar in both samples: people believe temperature has changed, and will change, much more than the estimates provided by the Intergovernmental Panel on Climate Change (IPCC). These beliefs are very unrealistic,

[3]Presentation is available at http://www7.nationalacademies.org/hdgc/Mental%20Models%20of%20Climate%20Change%20Daniel%20Read%20Yale%20University.pdf [accessed September 2010].

suggesting that if people fully understood the IPCC projections, they might lose interest in climate change as a problem.

The researchers tried to understand whether people distinguish between climate and weather. Between 1992 and 2009, knowledge declined on several questions about this topic. The researchers have not yet selected subsamples to equate on education. Responses to open-ended questions about what might cause global warming indicated not only many similarities across time, but also two important changes. In 2009, respondents were much less likely to attribute warming to aerosol cans, chlorofluorocarbons, and ozone depletion, or to loss of biomass. Read concluded that public understanding is very volatile and is possibly reflective of changes in media coverage. Results with closed-ended questions were similar. The rank ordering of causes is similar in both samples, but some items were considered less important in 2009 (e.g., deforestation, the hole in the ozone layer).

There were some changes in the effects that respondents anticipated from climate change. There was an increase in the extent to which respondents expected a number of impacts (ecological disasters, more frequent and larger storms, increased precipitation and humidity globally, and war and immigration problems), all of these consistent with scientific projections. However, there was decreased expectation of sea level rise and of shorter, milder winters globally—both changes in opinion that are opposite to scientific expectations.

An open-ended question about the most effective actions to help prevent global warming yielded the same most common response as in 1992—reduce driving. However, political action and raising awareness, which were the second and third most frequently mentioned responses in 1992, were mentioned much less frequently in 2009. The second and third most frequently mentioned responses in 2009 were recycling and saving energy, which were mentioned more frequently in 2009 than in 1992. There were some major changes in response to a parallel open-ended question on what are the most effective actions the U.S. government could take. Reducing auto emissions was at or near the top of the list in both samples, but protecting biomass, limiting pollution, and protecting the ozone layer were much less frequently mentioned in the 2009 sample, and alternative energy was much more frequently mentioned.

Closed-ended questioning about the effectiveness of various actions for preventing global warming indicated that actions that are generally seen as green are ranked highly, regardless of whether they in fact limit climate change, indicating a conflation of protecting the environment in general with preventing climate change. For example, stopping pollution from chemical plants, stopping the use of aerosol cans, recycling consumer goods, and compliance with the Clean Air Act were all among the more highly rated actions in both samples.

Read concluded that, overall, mental models have not changed a lot since 1992, despite a lot of publicity. There was less reference in 2009 to the burning issues of 1992, such as the "ozone hole," but some common misconceptions of 1992—the confusion of climate and weather and the pollution model of climate change—remained prevalent in 2009. There was some evidence of better understanding of the role of greenhouse gases.

Read spoke briefly about an interview study his group carried out in Seattle in 2008. It revealed a "natural causes" or "natural cycles" story about climate change that some respondents offered, which was not revealed in the survey study.

In the discussion following the presentation, a participant from the U.S. Environmental Protection Agency (EPA) reported that an EPA study has indicated that 42 percent of U.S. emissions are due to materials management, suggesting that recycling makes a huge difference, contrary to an implication of the presentation.

PERCEPTIONS OF CLIMATE CHANGE RISKS

Elke Weber[4]
Columbia University

Elke Weber began by noting that people study risk perception because of its importance in responding to hazards. It prepares organisms for action by changing physiological stress levels and affecting immune reactions; in addition, perceived risk combined with perceived control leads to positive or negative emotional reactions. This emotional aspect of risk perception serves as an early warning system that motivates action and also leads to expectations of actions by others.

Perceptions of climate change risks are influenced by cognitive and affective processes. One feature of climate change that has cognitive implications is the gradual nature of the change, which makes "signal" detection (i.e., noticing that there is a change) difficult. As a result of people's great adaptability, they sometimes do not even perceive gradual changes.

The uncertainty and time delay that are characteristic of climate change are additional cognitive challenges to taking protective action because the costs of climate change adaptation or mitigation are immediate and certain, but the future benefits of such action are uncertain and delayed in time, with large discounting as a result. Moreover, if such actions are successful in terms of preventing future negative consequences, the result may be that

[4]Presentation is available at http://www7.nationalacademies.org/hdgc/Insights%20from%20Research%20on%20Risk%20Perception%20Elke%20Weber%20Columbia%20University.pdf [accessed September 2010].

the actions appear to have been unnecessary. Weber noted that when people are considering possible future losses, they tend to be risk seeking—that is, they tend to take their chances. Thus, if climate change is seen in terms of future possible economic or environmental losses, there is a tendency to accept the risk and take the chance on future losses. She also noted that time discounting is not done at a constant rate per time period but follows a hyperbolic form (anything that does not occur right away is discounted very highly), and the discount rate is typically large. Although losses are discounted less than gains, time discounting rates are still very high even for losses and are higher than the rates typically advocated or used in current economic models.

Weber noted that how information is acquired matters. For example, many potential consequences of climate change are low-probability, high-consequence events. People tend to overweight rare events when they are described symbolically, so rare potential climate events could get a strong reaction. However, when people learn from personal experience, recent rare events are overweighted, but those that have not been experienced tend to be ignored. Rare events have a low probability of having occurred recently and thus will have low impact on perceived risk. Personal experience tends to be a stronger determinant of choice than vicarious descriptions because it is more engaging. Consequently, climate change events are not likely to provoke strong reactions in many people.

The absence of a visceral reaction matters, because emotions drive action. Analytic considerations are neither necessary nor sufficient for action. Evolution has prepared humans for simpler risks: "dread" and "unknown" risks get much stronger reactions than those that do not have these characteristics. Most people do not dread climate change. When affective reactions do exist, they are often incorrectly calibrated or misdirected. Climate change is not a risk people are hard-wired to care about. The threats are slow, intangible, uncertain, and statistical, lie in the future, and are not caused by a hostile agent. These characteristics help explain why global warming is low on people's policy priority lists. Poll data show that the importance given to climate change has dropped when there are other major worries (e.g., after the September 11, 2001, terrorist attacks and the recent economic downturn), indicating that people may have a finite pool of worry.

Emotions affect how people process information, with different presentations of information pressing different emotional buttons. For example, carbon offsets are more palatable than carbon taxes, especially among Republicans. Recent research elaborating "query theory" indicates that people argue with themselves, evaluate alternatives sequentially, and generate more arguments about the first-choice option they consider. Thus, whatever option is considered first gets more consideration. Default options (i.e., the option one gets if no decision is made, often the status quo) are so popular at least in part because they are the ones people consider first.

Weber noted several lines of theory and research that discuss risk as a social construction: the cultural theory of risk propounded by Douglas and Wildavsky (1982); the social amplification perspective (Pidgeon, Kasperson, and Slovic, 2003), which identifies as an important role of the mass media creating tipping points in public reaction; and research on how perceived risk can be affected by the expectation of action by others.

In final comments, Weber noted the mismatch between the magnitude of the problem and the nature of the solutions offered, as well as the absence of major events that might catalyze action. She said that Hurricane Katrina came closest, commenting that people need to be better prepared to respond to such events when they occur. Weber emphasized that perceptions of climate change risk are multiply determined, that nonanalytic processes are very important, and that affective reactions often guide cognitive processes. She argued for better appreciation that perceptions are malleable—that risk is not an immutable attribute of an event or action, but rather a judgment and a feeling that are constructed. Therefore how attention is focused, how information about action alternatives and their outcomes are acquired, what attributions are made about the causes of events, familiarity with events and outcomes, and perceptions of control all matter.

CLIMATE CHANGE DENIAL AND CONSERVATISM

Riley Dunlap
Oklahoma State University

Riley Dunlap began with the observation that the U.S. conservative movement has had a significant impact on debates over climate change. The historical background of this impact, Dunlap said, can be traced to the rise of a powerful conservative movement in the 1970s in reaction to what conservatives saw as threats posed by the progressive movements of the 1960s. The success of this "countermovement" can be seen in the general rightward shift of the U.S. policy agenda from the Reagan administration onward. After the fall of the Soviet Union in 1991 and the emergence of global environmentalism with the Earth Summit in 1992, the movement began to focus increasingly on the perceived threat posed by environmental regulations. It basically substituted a "green scare" for the declining "red scare." Conservatives launched an antienvironmental countermovement to combat environmentalism, which they saw as a threat to the conservative agenda of laissez-faire economics, free trade, the privatization of resources, and so forth.

During the Reagan administration, the movement learned that direct attacks on environmental regulations can produce a backlash and that it was more effective to question the seriousness of environmental problems. Because proponents of environmental regulations typically employ scientific evidence to make their case, the movement began to challenge such evidence as a key strategy. It did this by promoting "environmental skepticism," a dismissive view of the scientific evidence for environmental problems, particularly by "manufacturing uncertainty"—a strategy long employed by the tobacco industry and other industries to fight governmental regulations. Dunlap cited a book by David Michaels, *Doubt Is Their Product*, in which the author notes that "industry has learned that debating the science is much easier and more effective than debating the policy" (Michaels, 2008:xi).

Dunlap and his colleagues have been studying the efforts of the conservative movement to undermine the scientific evidence for environmental problems and climate change in particular. One study examined 141 English language books espousing environmental skepticism (Jacques, Dunlap, and Freeman, 2008): 81 percent of these were published after the 1992 Rio summit, and 92 percent were linked to a conservative think tank either by author's affiliation, publication by a conservative think tank press, or both. With the emergence of the 1997 Kyoto protocol, these books gave increasing attention to climate change, portraying efforts to limit global warming as threats to economic growth, free enterprise, personal freedom, and the American way of life.

In the past decade, climate change has become the preeminent environmental issue for conservatives, and manufacturing uncertainty has become the primary strategy for challenging the evidence for anthropogenic climate change. In fact, several figures in climate change denial were previously heavily involved in challenging the evidence concerning the harmful effects of tobacco smoke. The small number of contrarian scientists who have challenged mainstream climate science now have been augmented by a wide range of actors in the conservative movement. (Evidence for these conclusions appears in a paper published since the workshop: Dunlap and McCright, 2010.)

Dunlap said that as a way of examining the growth and diffusion of climate change skepticism/denial and the role of the conservative movement in promoting it, he and Peter Jacques are working on a study of books on this topic published through 2009. They are examining links between 84 books and conservative think tanks, using the criteria employed in their earlier study, as well as the major themes of the books and the credentials of their authors or editors. There was a sharp rise in publication of these books in 2007, which is continuing, and several of the books are bestsellers. A total of 64 (or 79 percent) of the books are linked to one or more conservative think tanks, and all but one of those not affiliated with a think tank have been published in the last decade. A growing proportion of the books are written by people who do not have natural science doctoral degrees, including several that are self-published, and these volumes are less likely to be affiliated with conservative think tanks. Eight children's books have also appeared in recent years giving reasons not to worry about climate change. While books espousing climate change denial were published primarily in the United States early on, they are increasingly appearing in the United Kingdom and other countries, particularly those affiliated with conservative think tanks.

The climate change denial books take issue with each of the major IPCC claims: that global warming is occurring and will continue, that human activities releasing greenhouse gas emissions are a major cause of the warming, that global warming produces harmful impacts on human and natural systems, and that a response is called for if harmful consequences are to be avoided. Nearly half of the 59 books published in the 2000s still question the warming trend, almost 90 percent challenge the attribution of climate change to human activities, and close to three-fourths are skeptical about negative impacts. There is also "delay skepticism," or the argument that there is no need to do anything now, even if climate change is occurring, and all but four of the books endorse this view. Dunlap argued that while the counterclaims to the IPCC have changed as the evidence supporting anthropogenic climate change accumulates, the bottom line in these books remains the same: no regulations. This reflects the near universal conservative ideology behind different versions of climate change denial. In fact, the sequence of arguments over time parallels that used in the past by contrarians (e.g., regarding the smoking-cancer link, acid rain).

Dunlap said that the counterclaims presented in these books and a wide range of other fora employed by contrarian scientists and conservative think tanks have been effective. He said that the U.S. media have given much more attention to climate change contrarian arguments and have been more likely to portray climate science as uncertain than have media in other countries. Not surprisingly, the U.S. public consistently expresses less concern about climate change than do the publics in other developed nations and is more likely to perceive a significant lack of consensus in the scientific community. The United States has yet to enact meaningful climate policy and has been an impediment to international policy making. Dunlap said that climate change contrarianism has become a core tenet in conservative policy circles and now has hegemonic status in the Republican Party, as evident in recent criticisms of any Republican politician who calls for action to deal with climate change.

Importantly, there is widespread evidence of a polarization of the U.S. public on the climate issue. Climate change denial has diffused to the general public, particularly to the conservative sector. Before the 1980s, views of environmental issues were only modestly associated with political ideology and weakly and, frequently insignificantly, with party preference. In the 1980s and 1990s, environmental issues gradually became more politicized, and political polarization began to increase. Today, party identification and political ideology are both good predictors of environmental positions on many issues. For example, trend data on opinions about "when the effects of global warming will begin to happen" (already happening . . . will never happen) showed a very modest increase since 2001 in the percentage of people saying that the effects of warming have already begun, but the partisan difference continually widened, with Democrats up, and Republicans slightly down (Dunlap and McCright, 2008; see Figure 1-2). Furthermore, among Democrats, belief in global warming increases with education and self-assessed understanding of the issue, but among Republicans, higher education or levels of understanding have little impact on such beliefs.

On the question of whether global warming is due more to human or natural causes, the trend since 2001 has been flat, but again there has been a growing divide between Republicans and Democrats, with the gap growing

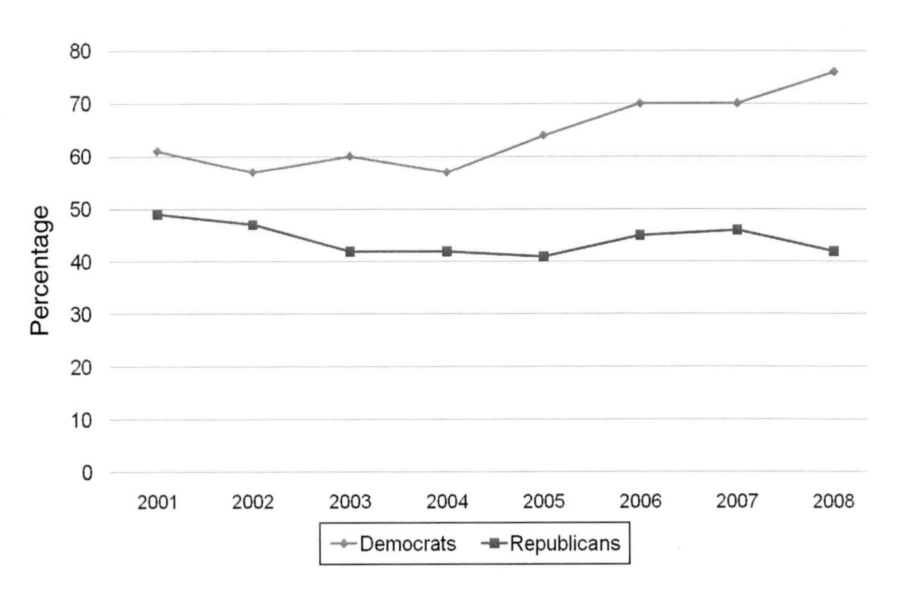

FIGURE 1-2 Percentages of Democrats and Republicans who believe the effects of global warming have already begun to happen, from Gallup poll data, 2001-2008.
SOURCE: Dunlap and McCright (2008, Figure 1). Reprinted with permission.

most among people who think they understand climate change very well. Dunlap summarized this evidence by saying that for a significant portion of the public, the conservative movement has been successful in portraying climate science as a hoax, a liberal plot, or "junk science" pursued by self-serving researchers.

Dunlap concluded by saying that, in addition to focusing on mental models and other cognitive phenomena impeding effective climate change communication, social scientists need to pay attention to the increasing flow of messages that are undercutting mainstream climate science. Lack of public acceptance of climate science does not occur by happenstance or stem predominantly from cognitive limitations; it is clearly affected by the perceived uncertainty concerning climate change that is purposefully and effectively generated by the climate change denial movement. The scientific community cannot craft more effective messages regarding climate change for the American public without taking into account what Dunlap described as the barrage of "disinformation" the public receives from those intent on undermining the credibility of climate science and thus the need for climate policy.

INVITED COMMENT

Frank Niepold
Climate Office, National Oceanic and Atmospheric Administration and Interagency Working Group on Education, U.S. Global Change Research Program

Frank Niepold said that the climate literacy/outreach/extension community has known for at least 2 years that knowing more climate science will not get the problem solved, that the "information deficit model" is not adequate. The education community is coming around to the social science work, and he expressed interest in making a greater effort toward understanding this work.

Mentioning the 2009 federal climate literacy document (U.S. Global Change Research Program, 2009a), he went on to say that education reform is hard to do in the United States because of state-level control. There are 15,000 school districts, plus museums and other venues for learning. In his view, a very broad consortium is needed to engage the country. Because it is easier to sow doubt than to remove it, people will be doing climate change education for a very long time. He said that to raise the literacy of the nation, both audience differentiation and sustained engagement are needed (which educators do naturally, but advertisers and people in communications do not).

Niepold called on the social science community to help the education community figure out how to use the social sciences in the education process. The disconnect between the two communities is a chronic problem, and he encouraged that community to help educators learn how to use social science work in education. He noted signs of increased attention to the topic. There has been a shift from an attitude of "let them know it's a problem" to "then what?" He said that teachers in some schools are doing 3-week units on climate change in fifth grade that leave some children crying. Many teachers do not have expertise in climate change science, nor in how to deal with the impacts of what they are teaching (i.e., the pupils' emotional responses). And although many are not familiar with solutions to the climate change problem, they need to teach about them in a way that will leave the students with some hope. Teachers need help to be fluent with the science and to teach what can be done about the problem. Niepold concluded by saying that federal science agencies, including the Agency for International Development, the National Institutes of Health, the National Oceanic and Atmospheric Administration, the U.S. Departments of Defense and Energy, and the U.S. Environmental Protection Agency, are working together on education, but they need help with best practices and techniques.

Roger Kasperson asked whether it is realistic to expect any major change in the American public's views. Niepold said it is, because federal science agencies are working hard on the issue and will monitor change, focusing on particular target groups. He noted that the United Nations (UN) Framework Convention on Climate Change requires all signatories to do outreach and said that not enough has been done in the United States in this respect. He said that this type of work is not sufficiently funded, and a way needs to be found to do it. Both federal agencies and nonfederal actors are increasingly aware of the issue.

In the discussion following the presentation, Dunlap noted a need to be sensitive to the climate contrarian literature moving into the classroom, for example, in the form of a clean coal coloring book given to schools for free or through the influence of conservatives on school curricula, textbook choices, and pressure on teachers. Stewart Cohen asked whether there is an information flow to professional networks—engineers, accountants, public health workers, among others—to engage them as professionals and turn them into extension agents. Niepold responded that the discussions of a National Climate Service (NCS) focus on this. Although it is not yet sufficiently resourced, there is talk about building capacity in professional associations. Cohen added that sustained conversations also have to occur in the professions. Niepold agreed that education in these communities would require a sustained effort. There was a question about involvement with the business community. Niepold said that in the NCS discussions, there was also emphasis on dialogue sessions with the business community to learn about its needs. He added that these discussions imply a paradigm shift in the science community toward public-private partnerships, in addition to the usual emphasis on satellites, physical observations, and climate models.

INVITED COMMENT

Bud Ward
Yale Forum on Climate Change and the Media

Bud Ward began by stating that the United States is in the early stages of a long and fundamental transformation of the media. He said journalism lacks a successful business model that supports quality journalism as a component of profitability. He noted that Ann Arbor, Michigan, now has no daily newspaper and predicted that other major U.S. cities also will find themselves without print dailies in coming years. There have been major layoffs at many newspapers, an ongoing process that he said has come to be known as "journicide." There also has been a reduction of specialty beats, with what had been environment specialty beats now expanded to cover issues as broad as environment, science, medicine, space, and technology. Ward said that the old rule of thumb of reporters spending 80 percent of their time doing reporting (research) and 20 percent of time doing writing has now been flipped, with reporters under intense and continual pressure to "feed the beast" of both their print and online outlets, including such electronic media as Twitter and Facebook.

Ward emphasized the importance of having climate change issues reported beyond the science and environment pages given that the issue can affect education, business, religion, travel, national security, and other news beats. Although such broader reporting is needed, it carries the risk of a return to an overemphasis on a false or simplistic "balance" or the inadvertent insertion of factual mistakes, because reporters new to the climate change issue lack familiarity with the scientific underpinnings. So, as the issue moves to other news beats, there will be a fallback in the learning curve. Science and environment journalists by and large accept that Earth is warming and that humans are significantly responsible, and they are being accused of being "template followers" by those critical of the scientific evidence. Another issue with news coverage is that if a story discusses uncertainty or risk assessment, it ends up being buried or killed outright. Pointing to increasing financial pressures facing the media and to something of a collapse of the traditional "iron wall" separating editorial and business interests, Ward said that the term "media industry" is now apropos. Today, journalists have to entertain as well as inform, and climate stories are not always amenable to this treatment. He agreed with Moser's observation that the media are not an educational institution. Reporters educate, he emphasized, but they are not educators, and the public should not expect the mass media or the mini-media to be educators. Ward anticipates that the revolution in mainstream media will result in a lot less investigative journalism. He concluded that how society will handle climate change will depend on how it handles journalism and other means of informing the public in a democratic system.

DISCUSSION

Much of the discussion focused on questions related to how greater levels of public concern and action might come to pass in response to climate change, considering the challenges to public understanding identified in the presentations. Andrews suggested that the conversation had evidenced an absence of positive ideas for framing climate change, for example, in terms of energy policy needs, economic development, and so forth. He noted that there are business allies for action on climate change. For example, the electric utility industry is seriously divided—Duke Energy has come out for a carbon tax, for example. He said there is a need to get beyond the information deficit idea and the environmentalist model in which public understanding and action all follow from rational science.

Moser suggested that it is important first to affirm the audience before presenting new ideas. People want to be "good people," so messages favoring climate responses should frame the responses as "what a good person does." She said that Americans want to see the positive side of everything and suggested that, in Europe, it is easier to have a conversation focused on difficult societal changes and on an important role for government in overseeing and guiding them. She suggested the value of framing messages in terms of how to be a good person while facing the realities of climate change.

Dunlap said that although positive framings, such as in terms of "green" jobs, are being made, public opinion has not changed. He noted that companies positioning themselves as green are vilified by conservative think tanks as anticapitalist.

Weber noted that people have multiple goals, including long-term and social goals, which may be activated as they make choices. Read pointed out that, in one study, telling people what a 9 cent/gallon tax on gasoline would be used for (to clean up the pollution caused by a gallon of gasoline) greatly increased willingness to pay.

Miron Straf noted methodological issues with survey responses, which are influenced by question wording, but identified ways to get past them. For example, people can be asked to think aloud about survey questions. Deliberative polling can also be used. Moser said that some research has gone beyond self-reported subjective opinions, but little has yet integrated deliberative processes with the goals of education. She suggested that if attitudes and beliefs are to change, different tactics must be used, including well-led deliberative processes. Dunlap said there have been some examples of such approaches.

Thomas Dietz pointed to some additional complexities in promoting greater public understanding. Recent analyses from the Stanford Energy Modeling Forum suggest that it is no longer possible to keep global warming below 2 degrees Celsius, "unless we continue to pump out aerosols." He said that because of the caveat about aerosols, it is incorrect to simply say that the 2 degree target has already been overshot. Moser suggested that no single message can adequately convey an understanding of climate change, noting the need to consider mitigation and adaptation together. She suggested that people would be more willing to hear such a complex message as part of a dialogue. Also, she emphasized that a message about the difficult challenges and great risks has to be paired with a message about the positive, constructive things that individuals, communities, and a nation can do. Without options for action, the conversation ends—environmental despair is a huge issue.

Nicholas Pidgeon noted that the science of climate change impacts is increasingly using the language of uncertainty. He asked how one can separate the uncertainty about impacts from the much lesser uncertainty about whether climate change is happening. He said that research in political psychology shows that conservatives reject any presentation of the issue that includes mention of uncertainty and suggested that this audience needs a framing that has more certainty. Dunlap noted that for a climate contrarian, even 95 percent confidence is evidence of uncertainty.

Cohen suggested that the climate deniers need to defend their certainty that climate change is not a danger and claimed that they are never asked to defend their beliefs. Ward added that climate scientist Stephen Schneider often said that scientists who speak with certainty are engaging in political rhetoric. Leiserowitz suggested that there are widely resonant frames available for talking about responses to climate change that address the uncertainty, such as making the analogy to buying insurance or gambling with the future.

Leiserowitz concluded the discussion by noting that a lot has been said about the complexities, the barriers to changing behavior, and how the problem is difficult. On the more positive side, he said that researchers have not really applied themselves scientifically to this question. He said that with an empirical approach, a lot can be learned. He noted that even if people understand climate change, they may not change all their relevant behaviors. Different behaviors present different barriers, and analysis has to become more sophisticated. He agreed with Andrews that there has been tremendous change in the corporate world, which has been responding to the huge financial opportunities in solving the climate problem. Cities, states, the federal government, many civic and environmental groups, and religious groups also are engaging. He concluded by saying that nature bats last: there will be teachable moments to which scientists will need to respond.

2

The Potential for Limiting Climate Change Through Household Action

Energy use in homes and in nonbusiness travel accounts for about 38 percent of U.S. carbon dioxide emissions (U.S. Energy Information Administration, 2008; Gardner and Stern, 2008). Household action could therefore make a considerable contribution to reducing national emissions. Much attention has been given to achieving those reductions by changing financial incentives through higher energy prices or subsidies for energy efficiency. However, significant potential for reduction exists even with existing economic incentives. It has been estimated that energy consumption in the residential sector could be reduced by 28 percent with current technology in ways that would provide positive net present value for the households (Granade et al., 2009). This finding indicates a major energy efficiency gap—one that might continue to exist even with stronger financial incentives.

In this chapter several leading behavioral and social scientists report on empirical research that helps clarify the basis for the energy efficiency gap and identify effective strategies for narrowing it. Although efforts to reduce household emissions have varied greatly in their effectiveness (National Research Council, 1985; Gardner and Stern, 2002), the most effective interventions show the potential impact. Research by Thomas Dietz and his colleagues shows that if the most effective documented methods were implemented nationally, the nation could achieve a reduction in household emissions of 20 percent within 10 years, without new technology and with little or no reduction in household well-being.

The presentations identified the key elements of the most effective programs, which use several instruments of behavioral change apart from or in addition to financial incentives. The presentations concluded that programs aimed at households are more effective when they feature strong social marketing, reduce the cognitive burden of making wise energy choices, provide for quality assurance, make information available from trusted sources at points of decision, and appeal carefully to existing social norms. The research by Charles Wilson suggests some additional program features that can be applied in programs to change household behavior by influencing manufacturers and retailers. Although much is yet to be learned, the behavioral research presented in this chapter increases understanding of the energy efficiency gap and identifies a variety of instruments of behavioral change that might be effective with respect to household action, in combination either with the current economic incentives or with enhanced ones.

INTRODUCTORY COMMENTS

Loren Lutzenhiser
Portland State University

Loren Lutzenhiser introduced the session by pointing out that this is a very timely topic. Local governments in climate-conscious areas are taking action, as are some states. For example, California is heavily engaged in policy development aimed at reducing greenhouse gas emissions, including regulatory requirements for buildings with zero net energy use in the commercial sector by 2020 and in the residential sector by 2030. The goals are very ambitious, but no one has a concrete idea how to accomplish them. In most of the climate policy framings, hardware changes are emphasized and users are absent from the analysis. Similarly, there is a significant amount of investment in "smart grid" activities, with billions of dollars of public and private funds being spent on new information and communication technology. Again, the investments are focused on devices, but in this arena there has been increased interest in providing better real-time feedback about consumption to energy users, to make more visible the energy flows that have long been invisible.

The social sciences have considered ways to promote behavioral change in many nonclimate policy areas, including health, education, and mental health. However, the policy goal has been to change the behavior of children or deviant groups. They have not previously tried to change the behavior of large segments of the adult population over decades-long time spans. In the energy and climate arena, a lot of policy activity is occurring now, but very little involves the social sciences or social scientists. Yet very good work over the past three decades has provided insights about how consumption is constituted and about how and why behavioral change happens or does not happen in this arena.

Lutzenhiser said that the workshop panel will explore some of those insights, going beyond the simple cost–benefit and return-on-investment frameworks of consumer choice that have oriented policy in the past. He noted that a lot of work on investments in energy efficiency has assumed that consumers are making "rational" economic decisions—a view that is not well supported by evidence. Several of the presentations will look at the evidence on actual determinants of those choices.

THE BEHAVIORAL WEDGE: THE NATIONAL POTENTIAL FOR EMISSIONS REDUCTION FROM HOUSEHOLD ACTION

Thomas Dietz[1]
Michigan State University

Thomas Dietz presented work that he did in collaboration with Gerald Gardner, Jonathan Gilligan, Paul Stern, and Michael Vandenbergh (Dietz et al., 2009). He emphasized five key points from this study: (1) effective analysis requires not only a hardware focus but also consideration of who makes decisions; (2) there is substantial potential for mitigation via the household sector; (3) to estimate this potential, one must make realistic estimates of behavioral plasticity (how much the behavior can change) as well as elasticity (how much emissions would change if the behavior changes); (4) realizing the potential of household action will require effective policy; and (5) effective policy must be evidence based, including social science evidence.

Who makes decisions? Traditionally, decisions are defined in terms of technologies rather than decision makers. Households make decisions about many technologies: those for in-home energy use, transportation technologies, and technologies that affect energy use indirectly. Their analysis considers only direct energy use in homes and travel. This underestimates total household impact; nevertheless, U.S. direct household actions account for 8 percent of total global emissions.

What can be achieved? The researchers examined 17 action types (32 actions), all of which involve employing existing technology and have zero or low cost or a high rate of financial return. They categorized five kinds of decisions: (1) weatherization, (2) household equipment (e.g., vehicles and appliances, all of which have numerous nonenergy characteristics that matter to consumers), (3) equipment maintenance, (4) equipment adjustments, and (5) daily use behaviors. These classifications were based on behavioral

[1]Presentation is available at url http://www7.nationalacademies.org/hdgc/National%20Potential%20for%20Emissions%20Reduction%20Thomas%20Dietz%20Michigan%20State%20Universit.pdf [accessed September 2010].

characteristics of the actions. They estimated the number of households already taking each action and corrected estimated emissions reductions for double-counting. They calculated the potential emissions reduction if all remaining households adopted each action. If all households took all these actions, U.S. household emissions would be cut by 37 percent, U.S. total emissions by 14 percent, and global emissions by 3 percent.

The logic of a behavioral analysis is to develop science-based estimates of plasticity: how much behavioral change can actually be achieved. To make such estimates, the researchers went to the literature on programs that have actually tried to get people to take these actions. They estimated what could be achieved nationally if the most effective known programs were implemented, acknowledging that the best existing programs may not be the best that could be devised. They multiplied the estimates of plasticity—what is realistic—by potential emissions to get reasonably achievable emissions reductions (see Figure 2-1). That estimate was a 20 percent reduction in U.S. household emissions in 10 years (without a cap-and-trade policy), which amounts to a 7.4 percent reduction in U.S. emissions, or a 1.6 percent reduction in global carbon emissions. In terms of the "stabilization

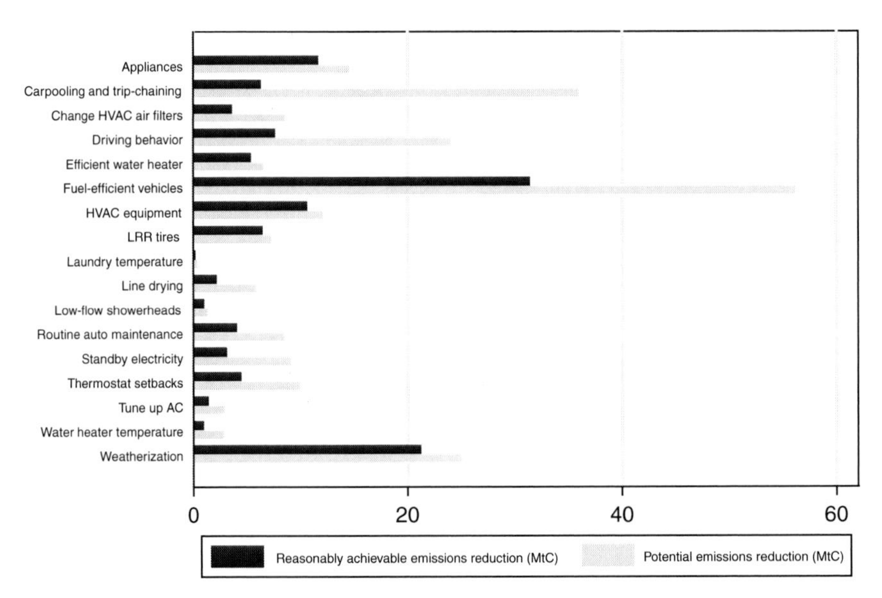

FIGURE 2-1 Technical potential and reasonably achievable emissions reductions from 17 household actions in the United States in million tons of carbon (MtC).
NOTES: HVAC = heating, ventilating, air conditioning; LRR = low-rolling resistance.
SOURCE: Data from Dietz et al. (2009). Used with permssion.

wedges" discussed by Pacala and Socolow (2004), this is the U.S. share of three wedges. And a 5 percent reduction in U.S. emissions could be achieved in 5 years with off-the-shelf technology.

Effective policy is essential to achieve this potential. This implies some research needs: to improve understanding of the current penetration of technology and practices; to improve data collection at the household level that attends to the needs for both physical/engineering science and social science analysis. There is a need to improve understanding of plasticity by developing a database for secondary analysis, using policies as experiments to improve understanding, and to conduct designed experiments at near-operational scale.

Dietz said that he and his colleagues were starting a paper on implementing the behavioral wedge (This work, Stern et al., 2010, and Vandenbergh et al., 2010, were published after the workshop.) Their reading of the evidence indicates that effective policies combine six program features: financial incentives (when the costs are nontrivial); strong marketing, including social marketing; information on how to take advantage of a program; convenience; quality assurance; and a focus on actions with high potential emissions reductions. The exact design has to be sensitive to the target behavior and the choice context. Dietz presented an exercise that rated three policies ("cash for clunkers," energy efficiency tax credits, and incentives for residential photovoltaic energy production) and gave them letter grades on the six features, which showed considerable disparity across the programs (much of this analysis is now available in Vandenbergh et al., 2010). He noted the lack of information on how well these policies have worked. He also pointed out that the incentives for photovoltaics vary from state to state, a natural experiment that could provide valuable knowledge.

Dietz concluded by suggesting some programs for weatherization that would score well on the program features mentioned. Offering a 50 percent incentive with the other program features in place, he said, could get 90 percent of the weatherization needs implemented in U.S. households. He also suggested the potential of "invisible" loans, the cost of which could be paid for out of savings on monthly utility bills. He recommended requiring posting of the energy cost of home occupancy in real estate transactions. Finally, he noted that Davis, California, has since the 1970s required upgrades of homes to city standards before sale.

In the discussion following the presentation, a participant asked about the potential for changing building standards. Dietz replied that the key is to find the leverage point, because of resistances in the building trades. Lutzenhiser said that the national laboratories have done a lot of work on this potential.

Another question concerned the split incentives problem between owners and occupants of rented space. Paul Stern commented that disclosure laws have some potential to address this problem. Susanne Moser said that it is too late if the disclosure requirement applies only at the time of transfer and suggested that it has to operate earlier in the decision process.

A questioner asked about change in the housing market, with developers not wanting to build the last "brown" development. Dietz noted that Edward Vine, in his Ph.D. dissertation at the University of California, Davis, in the 1970s, found that some developers want to do what they have always done, while others want to try what is new, to expand their markets.

INDUCING HOUSEHOLD INVESTMENTS IN ENERGY EFFICIENCY

Karen Ehrhardt-Martinez[2]
American Council for an Energy-Efficient Economy

The comments of Karen Ehrhardt-Martinez were based on preliminary insights from work to be commissioned by the Building Technology Program at the U.S. Department of Energy. She is putting together a database on what works, drawing insight from many sources, including the experiences of utility companies (much of it anecdotal), the proceedings of the American Council for an Energy-Efficient Economy (ACEEE) Summer Studies, and other sources. This effort supports four points.

First, there is widespread public support for energy efficiency, in contrast to the polarization of views on climate change. Ehrhardt-Martinez reported that polls show that about 78 percent of people say they should be spending thousands of dollars to make their homes more energy efficient, although only 2 percent report having actually done so. However, people do not have a clear understanding about what are the best methods of saving energy and making their homes more energy efficient. She mentioned a McKinsey report, indicating that only 15 percent of Americans see insulation as the preferred means to reduce their greenhouse gas emissions, compared with 50 percent who cited recycling and energy-efficient appliances. In another study, three-quarters of respondents estimated that home energy retrofits would save 10 percent or less, while savings estimates from professional energy analysts are in the range of 10-25 percent. According to a study by Dietz et al. (2009), the higher cost investments, which are the focus of this presentation, are likely to result in a 14 percent reduction in household energy use, out of the total estimated potential energy savings of 20 percent. She identified a broad range of drivers of consumption, citing a

[2]Presentation is available at url http://www7.nationalacademies.org/hdgc/Residential%20Energy%20Efficiency%20Karen%20Ehrhardt-Martine%20ACEEE.pdf [accessed September 2010].

report by a task force of the American Psychological Association (Swim et al., 2009). She noted that most policy has focused on technology, regulations, and income, but not on the other factors.

Second, regarding which kinds of retrofit interventions are most successful, she cited a National Research Council (1985) study indicating that grants and rebates were most attractive, followed by loans, but that higher income households favored loans. The study indicated that the size of an incentive does not determine whether people become engaged with a program, but it is important once people are engaged by affecting follow-through. The immediacy of the reward also matters. Other factors that matter include word-of-mouth communication and audits by trusted local groups (which were four times as effective in inducing action as audits by utility personnel).

Third, she noted that energy savings from feedback vary dramatically. Past feedback studies indicate a savings range of 0 to 32 percent, although most programs have experienced savings in the range of 5-15 percent. She is currently engaged in doing a meta-review. (This review, Ehrhardt-Martinez et al., 2010, was published by ACEEE after the workshop.) Although energy savings from feedback have been fairly well documented, there continues to be limited information about the behaviors that households are engaging in to achieve those savings. Results from the ACEEE review of feedback programs indicate that higher income households may be more likely to achieve home energy savings through investments in energy-efficient equipment, whereas lower income households seem to achieve savings through a shift in everyday practices that include low-cost or no-cost actions.

Fourth, she discussed evidence on barriers to purchases of energy-efficient appliances. A 1982 study of refrigerator purchases found that energy information at the point of sale is important, that labels were not very effective, and that emphasis by appliance sales staff on energy efficiency made little difference. Notably, however, the commission-based compensation structure for appliance sales personnel appears to lead to sales of more expensive and larger models of refrigerators, which, while efficient, tend to use more energy than smaller models.

Ehrhardt-Martinez summarized by saying that the most effective interventions typically target a particular action, address multiple barriers using multiple approaches, use social marketing methods and trusted information sources, and provide for convenience and quality assurance. The McKinsey study (Granade et al., 2009) identified barriers of awareness, the home ownership transfer barrier (i.e., the concern of homeowners that they will leave the home before the investment pays back), an inability to pursue savings, decision-making costs, and risk and uncertainty associated with contractors. Loans tied to property rather than to the homeowner could be effective at reducing some barriers. The McKinsey study also suggests that barriers could be overcome through the implementation of building performance mandates at the point of sale or during major renovations, as well as through the development of a certified contractor workforce. Ehrhardt-Martinez concluded by recommending the systematic collection of social science data at the national or state level, with data used to identify target behaviors, populations, barriers to adoption, and specific program strategies.

In the discussion following the presentation, Roger Kasperson said that a rational decision model underlies a lot of these presentations and wondered about the implication if that model is wrong. Are there other models that might work better? For example, diffusion theory suggests that the most important influence is whether a friend or neighbor has taken an action. That framework might supplement the analysis of investments. Ehrhardt-Martinez mentioned other social science models that have been applied: social norms research and concepts of identity and status as they relate to automobile choices. Kasperson noted that information might be most useful with early adopters and irrelevant for late adopters. Lutzenhiser noted that there are also potential applications of lifestyle theory and actor network theory.

INDUCING ACTION THROUGH SOCIAL NORMS

Wesley Schultz[3]
California State University, San Marcos

Wesley Schultz began by emphasizing that behaviors cluster and that they diffuse in nonrandom patterns, even within communities. Social norms—an individual's beliefs about the common and accepted behavior in a specific situation—are formed primarily through social interactions and exert powerful influence on behavior, especially in novel situations. He distinguished between injunctive norms (beliefs about what others approve of) and descriptive norms (beliefs about what other people do). Beliefs correspond with descriptive norms in self-reported energy conservation, for which correlation coefficients of .38 have been reported, and in other arenas. He reported on three studies.

After the 2000 California energy crisis, his group surveyed Californians to find the main reasons for conservation, which were self-interest, environmental concern, and social responsibility (Nolan et al., 2008). In an experiment, the research group provided materials that referred to these

[3]Presentation is available at url http://www7.nationalacademies.org/hdgc/Inducing%20Action%20Through%20Social%20Norms%20Wesley%20Schultz%20California%20State%20University.pdf [accessed September 2010].

three stated reasons, along with a social descriptive norm manipulation that stated a percentage of neighbors who do particular things to save energy. The group read utility meters for 4 weeks and interviewed residents in 1,207 households. The descriptive norm manipulation reduced electricity use by about 8 percent as compared with merely providing information about household electricity consumption. Appeals to the stated reasons for conservation had no effect. People said that the social responsibility and environmental messages motivated them, but that the descriptive norm was not motivational. However, the data showed a different pattern. Schultz emphasized that this manipulation affected relatively private behavior. He also noted that descriptive norm messages could serve as an anchor for people who conserve more than the average neighbor. He also pointed out that awareness campaigns typically implement a descriptive norm intervention incorrectly by saying that most people are not doing enough, which suggests a descriptive norm of inaction.

In a second study (Schultz et al., 2007), the research group gave people feedback about their energy use, as well as information about what the average home consumes. He suggested that this manipulation might get people to converge on the average from both sides. The study also added an injunctive norm: half the participants were given a handwritten happy or sad face emoticon depending on whether their consumption was less or more than the descriptive norm. The high consumers decreased usage after 2 weeks with the descriptive norm and decreased even more with the emoticon added. Low consumers increased usage when given only descriptive norm information but, when the injunctive norm was added, decreased usage slightly.

Schultz's group then worked with the firm OPOWER to turn the idea into a product, marketed to utilities, to be scaled up from the small experiment level. OPOWER does messaging through the mail, via bill inserts. A randomized control trial in Sacramento collected meter data for 50,000 households over one year: 35,000 received feedback plus an injunctive norm message from OPOWER by regular mail. Over 6 months, a 2.5 percent reduction in consumption was observed in this group. Households that had previously set conservation goals saved 8 percent. Households have been followed for 12-18 months in some utility service areas. The reductions do not diminish and, if anything, get stronger (see Figure 2-2).

In conclusion, Schultz said that social norms can play an important role in reducing energy consumption. More than feedback on energy use is needed, however. In particular, "smart" meters will not be sufficient. People need more than information. Normative messages can be scaled up and can provide an important supplement to information alone. In the discussion, a questioner asked whether information on cost savings reduces the effects of other framings of energy savings. Schultz replied that there is some

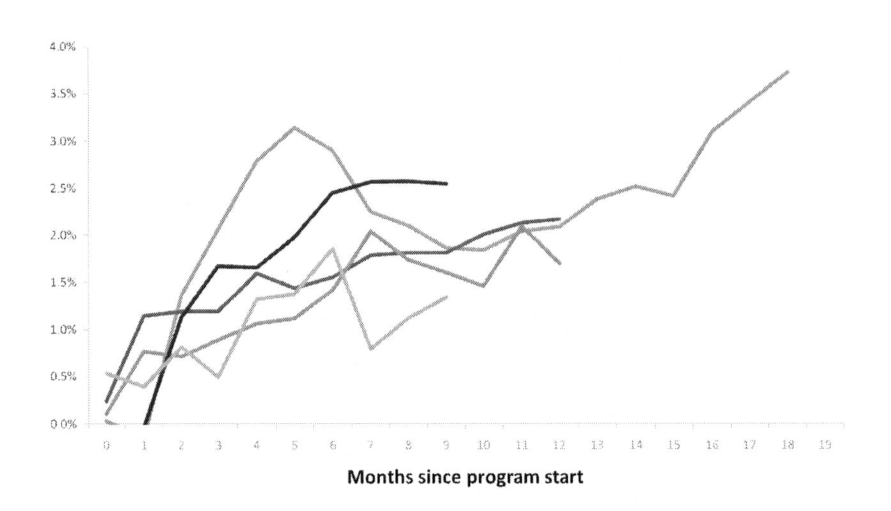

FIGURE 2-2 Residential energy savings over time (in percentage) in the OPOWER system as deployed by five utility companies.
SOURCE: OPOWER Home Energy Reporting system, see http://www.opower.com/Results/Overview.aspx [accessed September 2010]. Used with permssion.

evidence that framing energy savings in terms of money does make people think about energy that way, with the result that, if the savings are small, they may decide not to change their behavior.

Another questioner, saying that a pilot study indicated that using two frames produced a greater effect than one, asked if Schultz had combined social norms and public commitment. Schultz replied that commitment was a core issue in the old feedback studies, but that his research showed that norms can move behavior even without feedback or commitment.

In response to a question about personality differences in responsiveness to social norms, Schultz said that his group had failed to find moderators of the effect of norms, although they found equal levels of behavior change across countries. He noted, however, that Europeans say that norms do not affect them, while Chinese say they do. Finally, in response to a question about whether social norms can be used to increase actions that few people are taking, Schultz responded that injunctive norms might be useful.

SUPPLY CHAIN CONTRIBUTIONS TO THE HOUSEHOLD WEDGE

Charles Wilson[4]
London School of Economics

Charles Wilson explained that interventions in the supply chain are often referred to as market transformation because the intervention is in the market rather than with the consumer. The goal is the same: increasing adoption of efficient products, services, and practices. Interventions in supply chains are intended to create fertile ground for such policies as codes and standards.

Supply chains vary by technology and behavior. For appliances, the chain consists of manufacturers, retailers, and users. For home retrofits, the chain is longer. Chains tend to be more concentrated at the top and diffuse at the bottom. For example, there may be 5 glass manufacturers, 4,000 producers of window products, a larger number of installers, and 130 million homes. So there are fewer actors to change as one moves up the chain. Also, at different points in a supply chain, different interventions are appropriate. Training, quality assurance, and comarketing (e.g., ENERGY STAR labeling) all operate at mid-levels in the chain. There have been many successful initiatives, generally at the state level, to intervene in supply chains for energy efficiency. The American Council for an Energy-Efficient Economy disseminates these examples.

Wilson offered examples of best practice in market transformation. In Vermont, homeowners ask for energy audits, a request that triggers a process in which the recommended improvements are announced among certified contractors who can bid on the job, and financing is arranged for low-interest loans, with the contractor doing the paperwork. Financial incentives are provided for both contractors and homeowners. The incentive is set at 60 percent for rental homes and 33 percent for owner-occupied properties, but it can be greater if more retrofits are undertaken. This kind of program can provide incentives for retiring some equipment before the end of its natural life. In New York, a market transformation program targets appliances by providing consumer education about ENERGY STAR, placing information in retail stores, and training sales staff in order to create a constituency that can influence manufacturers' decisions about which appliances to produce and market.

Wilson offered a set of criteria for effective supply chain measures, saying that, for greatest effectiveness, programs should combine many types of interventions.

---

[4]Presentation is available at url http://www7.nationalacademies.org/hdgc/Interventions%20in%20the%20Supply%20Chain%20Charles%20Wilson%20London%20School%20of%20Economics.pdf [accessed September 2010].

1. Concentration—focus on a small number of actors for maximum impact, such as manufacturers, high-consuming segments of a population, low-income homes if they are being offered incentives, or entire neighborhoods.
2. Circumvention—seek ways to induce energy efficient behavior by reducing the cognitive burden of making well-informed energy efficient decisions.
3. Certification—this approach can provide quality assurance and establish trusted brands.
4. Changeable—whatever is tried should be adaptable. For example, a program to promote energy-efficient air conditioning on Long Island, New York, started by offering rebates and then shifted its focus to quality assurance, with sliding incentives that paid more for more efficient products.
5. Conditionality—make some program elements require others. For example, incentives for energy-efficient equipment could be linked to postinstallation diagnostic checks, to a program that provides data to manufacturers to help them improve their equipment, or to a behavioral commitment.
6. Contact points—focus on the points at which the household already comes in contact with the supply chain (e.g., retailers, home improvement contractors).
7. Cross-selling—use contact points to sell more energy efficiency than the program's initial target. For example, a program for home heating and cooling contractors in Texas gives the contractors additional incentives if they can get the homeowner to do a whole-house retrofit.

The measures in the Dietz et al. (2009) behavioral wedge paper can be divided according to the form of household contact with supply chain; 53 percent of the potential emissions estimated in that paper require direct household contact with the supply chain and can be influenced through the supply chain. These can be used to induce additional action (e.g., auto mechanics could check tire inflation while they tune engines). Some actions can involve indirect or incidental contacts with the supply chain and create different opportunities for influence (e.g., plumbers could train households to reset their water heater temperatures). Other actions involve no contact with the supply chain. However, actions that account for almost 70 percent of the potential emissions reduction estimated by Dietz et al. (2009) could be accessed through the supply chain.

Supply chain contacts can involve purchase, maintenance, and adjustment actions. One-time purchase or adjustment decisions can be influenced through contractors and retailers. Maintenance decisions might be influenced differently.

Wilson directed attention to the steady growth in homeowners' expenses for home improvement. Energy-related improvements account for only about one-fifth of the total, with repairs after disasters or to replace old roofing, siding, and so forth accounting for about the same amount (see Figure 2-3). The rest is for improved amenities. The home improvement supply chain is much larger than the energy equipment supply chain. He noted that homes have different meanings. Households may see the home as a haven, as a project, and as a social space (an arena for activities), and these meanings imply different motives for action. Home as a project is driving many nonenergy expenditures. He noted that people's decisions about home amenities may be different in important ways from their decisions about energy improvements.

People's stated motivations for amenity improvements tend to be emotional before the decision and more cognitive afterward. Wilson noted that the scope of home improvements often expands during the decision process, as households commit to going ahead with the improvements, confirm their affordability, and become more informed through contact with the supply

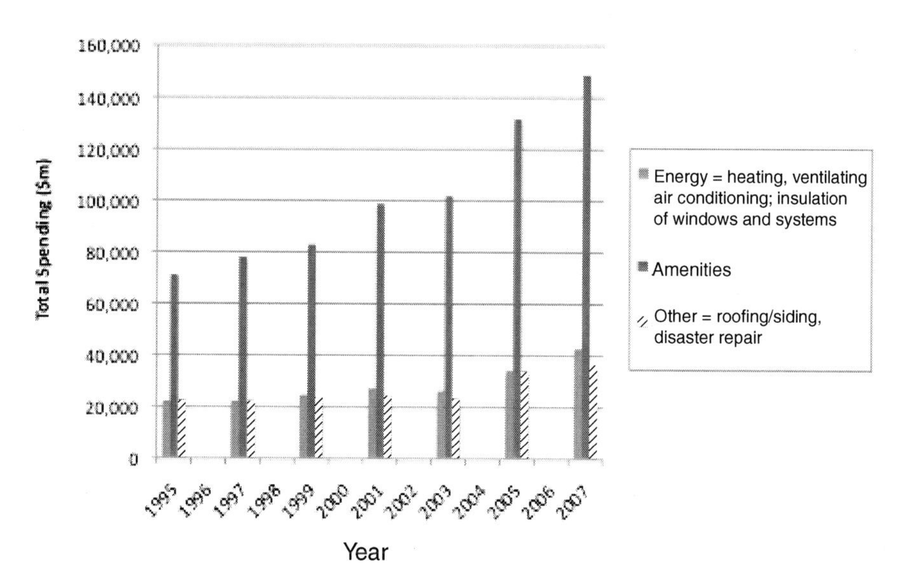

FIGURE 2-3 Homeowner expenditures for home improvements, by purpose, United States, 1995-2007.
SOURCE: Charles Wilson. Used with permission. Data are from tabulations of the 1995-2007 American Housing Survey by the Harvard University Joint Center for Housing Studies.

chain. This fact provides opportunities to bundle energy efficiency improvements with amenity improvements. Doing this can help people rationalize their amenity renovations. However, he noted that the supply chains for these two kinds of renovations are largely different at present.

Wilson concluded by emphasizing the need to try what one thinks will work based on what is known about how households actually make home improvement decisions. Two key elements are reducing the cognitive burden of informed decision making and leveraging existing contact points between households and the supply chain. As an example of the latter, he noted the opportunity for market transformation programs to work with real estate agents, for example, by requiring that homes being sold have an energy performance label, as is increasingly required in Europe. However, he noted that the real estate industry has not been used much as a channel for promoting energy investments.

A questioner noted that many people in the United States see their homes as an asset and asked about the implications of that view. Wilson said that there is evidence that the value of renovations correlates with home prices. He found in his research, however, that home appreciation is mainly a post hoc rationale for expenditures.

A related question concerned how to make energy improvements more visible. With an amenity improvement, a real estate agent might recommend that the seller fix the kitchen and then ensure that potential buyers look at the kitchen, thus making the investment both visible and valued. Could real estate agents be educated similarly to ensure that energy efficiency is valued by the marketplace? In response, Wilson noted that some jurisdictions have green training and certification programs for real estate agents, as well as certification programs for homes at the point of sale.

DISCUSSION

Lutzenhiser opened a general discussion by asking how these insights could be applied to policy. He said that these presentations have shown the value of a social science look at the programs and possibilities for household action. However, he noted a mismatch of the social science and natural science with policy making and implementation. He said that a program cannot be executed from theory. The best programs can be described in terms like those used by Wilson, but only after the fact. Program design is not a matter of applying social theory. Because of the diversity of behaviors, technologies, and decision contexts, one size does not fit all. Variability is vast even in single-family homes in the same geographical area. This diversity creates an almost intractable problem, making it easier for a policy maker to "throw a price at it." The work by Dietz et al. (2009) focuses on things that are relatively painless. But there is room for a broader behavioral contribution that also considers how elastic comfort preferences are. Even larger savings are possible from lifestyle changes, but the research on this has not been done. Also, these studies have not addressed the human role in shaping technology.

Dietz said that social scientists lack a partnership with engineers and other technical specialists to do life-cycle analysis for lifestyle change. He also commented that the largest uncertainty in the emissions scenarios used for climate change research is what society will do in the future. If scientists want to have better emissions scenarios for the sixth Intergovernmental Panel on Climate Change Assessment, is there an alternative to using price elasticity? Ehrhardt-Martinez said that it is necessary to map the diversity in motives and in ability to take action and then to develop policies that take this diversity into account.

Economic in Comparison with Other Social Science Analyses

Much of the discussion addressed the difference between economic analyses of energy efficiency, based on presumptions of "rationality" that emphasize the material incentives that affect choices, and the approaches in these presentations, which draw on behavioral research and presume a broader range of influences on choice. Schultz noted that the presentations shared a critique of "rational" actor models. However, he saw a diversity of views and no uniform voice. He suggested that because the rational actor models present a core message and these presentations do not, the latter are likely to be dismissed. Lutzenhiser said that each discipline "takes a little bite of the apple" and communicates a little with the other disciplines, but large parts of the problem area are not addressed. He noted that some research "drags the problem back into the disciplines."

Wilson agreed with Schultz that behaviorally informed policy suggestions did not have the clarity of what economists offer, which includes estimates of how much interventions will cost and how effective they will be. He said that the behavioral research message is that these interventions influence the price elasticity of demand, as Stern (1986) has noted. Ehrhardt-Martinez noted important insights that behavioral research can offer, such as that households can make an important contribution, that framing actions in terms of energy is less politically controversial than framing the same actions as responses to climate change, and that information and incentives alone are insufficient to change behavior. Dietz observed that there is a subgroup of economists who know their models are too simplistic and are open to that criticism, but that they want to be told how to do it better—for example, why price elasticity is sometimes low and sometimes high.

Moser said that the interdisciplinary social science analyses are not as scattered as Schultz stated and suggested that, with sufficient effort, they could have a unified message. Dietz added that this kind of problem-oriented meeting, with more than just a few social scientists interacting together, is really rare and much needed. He asked how to make more such meetings happen. Ehrhardt-Martinez said that the annual Behavior, Energy, and Climate Change (BECC) conference is geared to do that and to get social scientists together with practitioners and government officials. Lutzenhiser said that although BECC is good for exposure of social science ideas, it does not meet the need for in-depth interaction.

Policy Applications of Social Science Work

Another theme was the application of the kinds of work presented at the session. Stern said that energy policy makers are now reinventing insights from behavioral research, sometimes creatively. He posed the problems of getting them to apply the insights systematically and to learn systematically from policy interventions. He also noted the need to build on the possibilities of local actions and to study those actions. He suggested the possibility of creating a Wikipedia-style database, with people supplying information on their local programs. Lutzenhiser noted that current efforts to promote weatherization are natural experiments that probably will not be evaluated closely, although social science could turn attention to it. A participant suggested that some private businesses might share data on their efforts.

John Dernbach observed that current climate legislation—cap and trade, tax credits, cash for clunkers—makes a large number of behavioral assumptions. For example, the distribution of allowances to utilities makes assumptions about how they will use them. Legislation needs to be designed so that cost increases will yield behavioral pathways that lead people in the desired directions. The design of legislation should require consultation with behavioral scientists in the same way that it requires consultation on the science of climate change. There might be legislative proposals to create ways for consumers to follow paths to lower their costs that would increase chances for enacting the legislation.

Kasperson suggested that, although the work discussed in this session covers a domain of knowledge that should be incorporated into policy, it is not happening. He said that to get this work implemented in policy will take more than just summarizing knowledge and giving it to people in a workshop report. Moser said that the receiving end lacks people who know how to implement what social science research has shown. There needs to be a dialogue with people at the receiving end who could implement what is known.

Nicholas Pidgeon mentioned his recent paper (Spence and Pidgeon,

2009), which resulted from a discussion about what could be accomplished by lifestyle change, in the context of what is being assumed about behavioral change in many current models of future energy scenarios. The fourth IPCC assessment report in 2007 had only five footnotes on behavior or lifestyle. The IPCC adaptation report expressed medium agreement that lifestyle change can contribute—but there was no analysis behind this conclusion. There is a need to tie the IPCC community down to these claims and require social science input and also to get the modelers to include more social science in their models.

Potential for Larger Behavioral Changes

Anthony Leiserowitz returned to the issue of larger behavioral changes raised by Lutzenhiser. He noted that society has multiple path-dependent systems and suggested that, in the big picture, the greatest leverage might come from life-changing moments that set new patterns. For example, there is huge potential for avoided emissions in developing countries where many people want to adopt U.S. lifestyles. Can one imagine a behavioral science approach to helping developing countries leapfrog the U.S. lifestyle and get more quickly to a lower emission future one? Ehrhardt-Martinez noted that there is a large potential if people "break out of the mold" of current lifestyles. Unlike Dietz's work, her study found that 57 percent of potential savings in emissions came from changed habits, routines, and low-cost actions. She noted in particular the case of Juneau, Alaska, which cut usage 30 percent in 6 weeks during a crisis.

3

Public Acceptance of Energy Technologies

The history of the U.S. nuclear power industry demonstrates that public concerns can derail the implementation of energy technologies even when many technical experts believe them to be safe, effective, and economical. The implementation process in that industry involved a large expenditure of effort, money, and time, but resulted in much less energy production than originally anticipated. Some observers believe it also resulted in a long-standing mistrust of government and the industry. Now, as rapid expansion of domestic energy sources has become a major national policy objective and as public opposition is appearing to a wide variety of energy developments—from wind farms to natural gas drilling to carbon capture and storage (CCS)—it is useful to look for ways to avoid a repeat of the nuclear power history.

In this chapter several leading scholars review the research on public reactions to newly emerging technologies, to the siting of potentially hazardous facilities, and to two specific energy technologies: offshore wind and CCS. The presentations showed that empirical research can help decision processes by identifying public concerns that are not otherwise obvious. Also, as the presentations indicated, the research reveals some recurrent themes, such as the importance of trust in the organizations that are promoting the technology and the need for two-way communication and the leadership and staff to follow through with it. The presentations also identified a variety of unanswered empirical questions about the most effective processes for identifying and addressing public concerns.

INTRODUCTORY COMMENTS

Roger Kasperson
Clark University

Roger Kasperson opened the session by noting that some years ago, a workshop like this one (organized by the International Human Dimensions Program on Global Environmental Change) concluded that any major restructuring of the U.S. energy system would require major social change. He noted that discussions of the development of low-carbon energy systems seem to be proceeding with thinking largely restricted to technological questions and with little recognition of the history of nuclear power. Development of the new low-carbon technologies is at the first stage of a process, with little attention yet to social science issues. He expressed pessimism about learning from experience, predicting that there will be very little interest in public acceptance until policy makers have been forced to face the experience of substantial public opposition. In his view, it will be hard to proceed with renewable energy developments without taking public acceptance issues into consideration.

Kasperson's views were shaped by a couple of decades of work on nuclear power issues. Current analyses of wind technology proceed with concerns focused on a single issue (e.g., threats to birds and bats), with limited attention to other issues (e.g., visibility, community concerns). In the 1970s, with nuclear power, the technical community similarly focused heavily on reactor accidents and paid scarcely any attention to nuclear waste. The waste issue was missed because analysts were convinced that they knew what the key issues were. He sees the same thing happening with the new renewable technologies, Kasperson said, although he expressed hope that he is wrong.

LESSONS FROM THE PAST: GOVERNANCE OF EMERGING TECHNOLOGIES

Nicholas Pidgeon[1]
Cardiff University

Nicholas Pidgeon began by stating that, although climate change has important social drivers, the solutions being offered are primarily technological and economic. These will not succeed without some degree of public buy-in and acceptance. His comments focus on large-scale technologies, al-

[1]Presentation is available at [url] http://www7.nationalacademies.org/hdgc/Governance%20of%20Emerging%20Energy%20Ttechnologies%20Nicholas%20Pidgeon%20Unidersity%20of%20Cardif.pdf [accessed September 2010].

though public acceptance issues also exist with small-scale ones. He pointed to the recent report of the National Research Council (NRC) *America's Energy Future* (National Research Council, 2009), which provides a list of technological possibilities. For these technologies, he said, a key issue in governance is differences between formal and lay understandings of risk. There are six issues, each with a lesson:

1. *Risk has qualitative characteristics.* Lay publics respond to more than the probability and consequence aspects of risk. Risk communication has to be a dialogue-based process if analysts are to speak to people's concerns.
2. *Cultural and institutional discourse matters*, which implies that values matter. There are many ways to "frame" the climate change issue or a new technology. There is no single public perception of risk. Choices can be framed as being about the environment, money, social movements, or global security. Pidgeon posed this question: Do the value differences about climate change drive disagreements about energy options? He cited data from a new study (Spence et al., 2010) indicating that concern about climate change is positively correlated with support for renewable energy, such as solar and wind, but not with support for building new nuclear power plants, even though nuclear power is largely carbon free (see Figure 3-1).
3. *Social amplification of risk.* Technological controversy is a dynamic social process that cannot be readily predicted or "managed." Participants in controversies try to influence each other. In the British controversy about genetically modified foods, there were arguments over many issues, including equity, trade, and distrust in food regulation, as well as about the technology. Public beliefs were ambivalent on the topic, but the public questions were about social issues, such as who will regulate, the balance of benefits, and a more polarized, and more strongly opposed, set of views than was present in a representative sample of the British public (Pidgeon et al., 2005).
4. *Risk and trust.* The importance of trust implies that openness, transparency, and dialogue are important, alongside responsible risk management. Social agreement, structural attributes of responsible agencies, and emotions are all related to trust. A recent British study shows that support for new construction at two existing nuclear power sites was strongly related to place attachment and to trust in the nuclear industry and much more weakly related

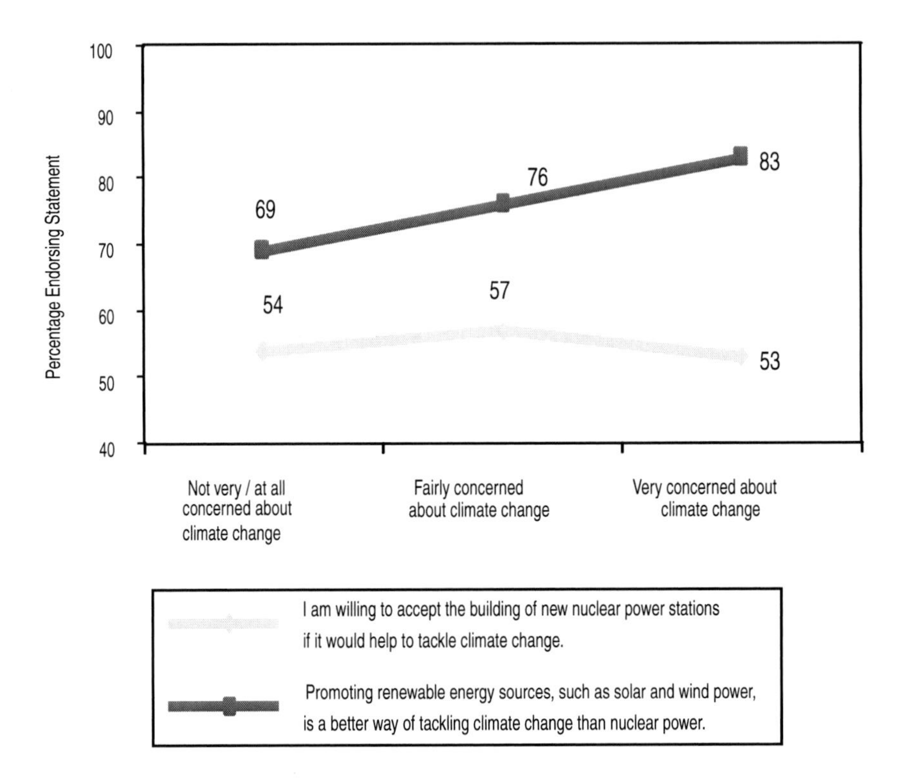

FIGURE 3-1 Attitudes to nuclear power and alternatives as a function of level of concern about climate change (% replying "agree" or "strongly agree") (n = 1,491, United Kingdom, surveyed in 2005).
SOURCE: Spence et al. (2010). Used with permssion.

to perceived benefits, perceived risks, and concern about climate change (Pidgeon et al., 2008).

5. *Properties of emerging technologies.* Emerging technologies present deep forms of uncertainty and complexity, a fact that implies a need for innovative modes of risk governance in addition to conventional risk assessment. Knowledge about outcomes is different from knowledge about probabilities, and either can be more or less problematic. As knowledge about likelihoods becomes more problematic, decision makers respond with heuristics for coping with uncertainty, by using sensitivity analysis, and by creating more reflective institutions. As knowledge about outcomes becomes more problematic, decision makers have to rely more on scenarios; backcasting methods; and processes that include participation, de-

liberation, and accountability. As both kinds of knowledge become more problematic, reflexive governance; monitoring and surveillance; and strategies emphasizing flexibility, learning, diversity, and adaptation become more attractive. Emerging risk perceptions for new technologies are hard to study because people lack direct experience, mental models are ill formed, there are inherent uncertainties about technology and its regulation, and there is hype and hope from technology promoters. Nanotechnology has some similarities to genetically modified organisms in this respect. The technology is at stage zero, making it hard to have any form of public debate about it. Lack of awareness is a major challenge because the organizers of dialogue inevitably have to provide a cognitive frame for the technology. Pidgeon emphasized the need to address fundamental questions, such as Why this technology? Who needs it? Who owns it? Who will take responsibility?

6. *Perceived benefits and use matter* both to lay people and to people who want to "amplify" the risks. The proponents need to demonstrate that they are producing a real public good. Pidgeon reported that his research group did a U.S.-UK dialogue on nanotechnology and found cross-national similarities. The main issues that arose were about choice, control, and uncertainties. The application of the technology (i.e., health versus energy) mattered a lot. People did not easily see risks from nanotechnology energy applications, whereas nanotechnology health applications were seen to raise distinctive ethical questions (Pidgeon et al., 2009). Pidgeon and colleagues concluded that there is a need to move away from sterile debate on whether or not to be precautionary and to develop new ways to assess risk and uncertainty.

In the brief discussion, Thomas Dietz asked whether it is known how people perceive the new industries and regulators that are promoting renewable energy technologies. Pidgeon replied that people generally have a very positive attitude toward wind and other renewables, which might give proponents a reservoir of trust. Susanne Moser asked whether Pidgeon's findings hold for a global intervention, such as climate geoengineering. Pidgeon replied that whereas one can organize a public participation process about a site, for a global issue it is very hard to do because the interested parties are global. He noted that there is an emerging critique of public participation around emerging risk issues that emphasizes the need to do participation differently, in an appropriate way for the decision context. Questions remain about the viability of public participation approaches for technologies on a global scale.

LESSONS FROM THE PAST: ADDRESSING FACILITY SITING CONTROVERSIES

Seth Tuler[2]
Social and Environmental Research Institute

Seth Tuler conducted a review of literature across several types of technology, emphasizing hazardous facilities and energy-related facilities and their associated infrastructure. He found almost no connection between the literatures on nuclear/hazardous and on renewable energy technologies. The literature is very large. Work on low-carbon energy sources is heavily skewed to wind, with very little recent work on siting solar arrays and only one study on geothermal energy. Studies use varied operational definitions of the key variable, which is variously defined as support, acceptance, tolerance, and success/failure. The studies are almost all site-specific or technology-specific.

Influences on support or opposition include anticipated outcomes on various dimensions (e.g., health, economy, sense of place, quality of life), and characteristics and preferences related to the planning and decision-making processes. Other mediating factors are also reported, including characteristics of the technology and its design, qualitative aspects of risk, issues of the credibility and competence of the managing institutions and the motivations of the developers, values, degree of exposure to hazards, and prior experience with the technology or its developer.

For hazardous facilities, worries about health and safety, risk dimensions, and hazard management dominate. For wind energy, the main factors are the transformation of the landscape and the place. Biomass and CCS facilities appear to be perceived somewhat like hazardous sites. Concerns about economic issues and quality of life play out differently with different technologies. Finally, perceptions of the local impacts tend to drive support more than concerns about national impacts, which are more of a focus in messages from technology supporters. This suggests that appealing to climate change might not be helpful in generating local support for a new energy facility.

With wind energy, the effect of proximity is complex. Studies show that after turbines are in place, proximity increases support, whereas the reverse is true before siting. This is probably true because the worries are about the landscape, and many such concerns can be settled fairly quickly after the turbines go up. With hazardous sites, the concerns are not alleviated by short-term experience. The research shows that general opinions about

---

[2]Presentation is available at [url] http://www7.nationalacademies.org/hdgc/Addressing%20Facility%20Siting%20Controversies%20Seth%20Tuler%20SERI.pdf [accessed September 2010].

a technology are not the same as opinions about a particular proposal; opinions are dynamic, and site-specific concerns are very important.

The literature offers many claims about how support and opposition change. One claim is that change in institutional frameworks can affect trust and the distribution of benefits. For example, in Denmark, support for wind energy decreased when the turbines were no longer owned by the communities. A second claim is that getting the process right is critical (National Research Council, 1996). But there are studies of places where there was support even among people who saw the process as unfair. A third claim is that providing information will increase support. Interventions may change opinions, but the direction of change is not given. There is some evidence that opponents are less likely to change their minds than supporters. Little is known, however, about the large reservoir of unengaged people.

Tuler summarized a few main points in this literature: that opinions are dynamic, that neither supporters nor opponents are all alike and that both can be ideological, that institutional frameworks and processes matter, and that context matters a lot. The key questions include what the right process and information are. There is not much in this literature to specify these for the new energy technologies now under consideration.

In the following discussion, Jeremy Firestone asked about research on power transmission sites. Tuler replied that there is very little research on the topic. He noted that one set of studies concluded that siting of renewable energy facilities has proved hardest in states with renewable portfolio standards.

Stewart Cohen noted that in the literature on climate change adaptation there is research on pathologies and research visioning for new adaptation technologies. He asked about similar research on energy technologies. Tuler said that there is a lot of experience with wind power, mostly in Europe, and some with CCS, mostly on visioning. Another participant asked whether people with experience from other wind sites have made a difference in acceptance. Tuler replied that some case studies assert that this does matter, but the evidence is mainly anecdotal.

There was a short discussion of how the findings might apply to climate geoengineering. There has been very little research to date. Kasperson commented that huge financial commitments are made in the new technologies despite very little understanding of public reactions.

There was also a brief discussion of state preemption of local involvement. Richard Andrews commented that in the 1980s with hazardous waste sites, the price of preemption to democracy was high and the benefits for siting were low.

ACCEPTANCE OF OFFSHORE WIND: GETTING TO YES ON A WEDGE OF A WEDGE

Jeremy Firestone[3]
University of Delaware

Jeremy Firestone framed his presentation in terms of the role of wind power in reaching the goal of an 80-90 percent reduction in U.S. carbon emissions by 2050. In a conventional view, most of the U.S. wind resource is in the Great Plains, so large transmission lines would have to be built to get this power to the East Coast. However, there are strong offshore winds on the coasts, mainly near metropolitan centers. The wind resource off the Atlantic coast can produce 58 gigawatts (GW) at locations with 20m or less of water depth, with 340 GW available if turbines can be sited at up to 100m depth. This could produce all the power needed for the region, which is currently 139 GW of generating capacity and 73 GW of average output. But it would require 54,000 turbines to generate 73 GW, which would require substantial public acceptance.

The turbines are large—up to 417 ft to the top of the blade—and so are the factories that produce them. Firestone's research group examined the public acceptance issues at two locations: Cape Cod and Delaware. The technology, while deployed off Europe for almost 20 years, is at stage zero or slightly beyond in U.S. waters, because there is no U.S. experience with offshore wind. In both study locations, much electric power now comes from a "dirty" plant (oil in Massachusetts, coal in Delaware). However, in Massachusetts, the existing power plant has not become an issue in the discussion over the offshore wind project, whereas in Delaware coal appears to be no longer socially acceptable. There are also differences in how people think about place. In Massachusetts, although the proposed site was beyond the 3-mile state limit, the site was in an area that was not considered open ocean because of nearby islands. The Delaware project is proposed to be 13 miles offshore, compared with 6 miles offshore in Massachusetts.

Firestone presented results of a survey conducted in 2009 on samples of people in Delaware and Massachusetts who live on the coast and farther inland. The researchers provided photo simulations of the view of the proposed site and told people to hold the page at the proper distance to get the sense of the view as it would appear at actual size. Support for siting was stronger in Delaware than in Massachusetts, and it had increased in both places compared with earlier surveys conducted in 2005 and 2006. In Delaware, even people who think they would see the project from

[3]Presentation is available at [url] http://www7.nationalacademies.org/hdgc/Offshore%20Wind%20Power%20Jeremy%20Firestone%20University%20of%20Delaware.pdf [accessed September 2010].

where they live were supportive, although this was not the case in Massachusetts (see Table 3-1). There have been 8 years of decision process in Massachusetts—something the developer did not want, but is now benefiting from, as support has been increasing. Many people there perceive the developer as transparent and the planning process fair. In Delaware, there is greater positive feeling about the process, although a larger proportion of people express no opinion.

Firestone reported that the main reasons for support are foreign oil dependence, especially in Massachusetts, and the possibility that wind power will mean stability in electricity rates. Strong majorities in both locations indicate they would be more supportive of their local project if it was the first project of some 300 offshore wind projects—that is, if it is part of a transformative energy policy.

In a 2006 survey in Delaware, the same researchers found that people prefer the turbines to be out of sight, but willingness to pay for moving them farther from shore is low beyond 6 miles. The survey indicated that an offshore wind turbine could move as close as 1 mile from shore before people would prefer a coal plant.

Firestone concluded that framing of the choices is important to public

TABLE 3-1 Support and Opposition to Offshore Wind Energy Projects in Delaware and Massachusetts Related to Expected Visibility of the Project from Their Homes

Visibility of Project		Delaware Ocean	Cape Cod Sound
Respondents Think They Will See Project	% of Stratum	12 percent	8 percent
	Support (%)	**69**	26
	Oppose (%)	**31**	74
	Unsure (%)	0	0
Unsure Whether They Will See Project	% of Stratum	11 percent	16 percent
	Support (%)	67	**55**
	Oppose (%)	27	**45**
	Unsure (%)	6	0

SOURCE: Jeremy Firestone. Used with permission.

acceptance. The choice is not between wind and nothing, but between wind and other energy production options. He said it is important to frame the choice in ways that get people to compare the options for meeting demand.

In the discussion, a participant asked if there is research on wave power. Firestone said that one study in Oregon shows public acceptance, but it is not clear if this indicates general or site-specific support. He noted that wave power is an issue on the West Coast because the continental shelf drops off quickly, making it difficult to site wind turbines at a distance using conventional technology.

Another participant asked if differences in acceptance at the two sites are due to the characteristics of the site (people pass through it in Nantucket Sound), the demographics and economics of the areas (e.g., Cape Cod is a high-income area), or the history, which made coal unacceptable in Delaware. Firestone replied that the first and third explanations resonate, but that the second does not. The place attachment of the coastal people is higher in Delaware. However, boating issues are relatively important in Massachusetts. He suggested that part of the difference may be that the Delaware culture values good government—people respond to the need to do the right thing.

John Dernbach asked if wind could be compared to Marcellus gas shale in terms of how fast it is to be brought online as an energy source for the East Coast. Firestone said offshore wind is unlikely to represent a significant fraction of electricity within the next 10 years, but it can be game-changing in the longer term for the East Coast. He noted that the Midwest wants transmission lines and suggested that they would be more controversial than offshore wind.

PUBLIC PERCEPTIONS OF CARBON CAPTURE AND STORAGE

Wändi Bruine de Bruin[4]
Carnegie Mellon University

Wändi Bruine de Bruin presented research that she did in collaboration with Lauren Fleishman and Granger Morgan, based mostly on Fleishman's dissertation. She began with the comment that, according to experts, CCS could reduce CO_2 emissions if the public accepts it. The few available studies suggest that people are lukewarm about the technology. However, the results are hard to interpret because the ratings were given after only a

[4]Presentation is available at [url] http://www7.nationalacademies.org/hdgc/Carbon%20Capture%20and%20Storage%20Wandi%20Bruine%20de%20Bruin%20Carnegie%20Mellon%20University.pdf [accessed September 2010].

very brief description and because there was not a comparison with other options. Effective communications are therefore needed to lead to more informed decisions.

Effective communications are based on input from experts to ensure balance and accuracy, as well as on formative research with the audience to ensure their understanding. The latter is usually not done. There have been few efforts to communicate about CCS, and they do not reflect best practice in risk communication. They have been developed without input from audience members and may therefore be ineffective or harmful. Her research group has tried to (1) elicit the audience's mental models of CCS in qualitative interviews, to identify relevant context and wording; (2) conduct quantitative surveys to compare interviewees' beliefs with those of a larger, representative sample; and (3) develop and evaluate risk communication materials with input from experts and nonexperts.

In the case of CCS, people began with too little knowledge to provide mental models, so the researchers had to provide some basic information before eliciting beliefs. Among a small group of interviewees, prevalent concerns were about negative side effects, efficacy, and costs. One-third referred to nuclear waste or "pollution." One-third wanted to compare CCS with other options, such as wind and solar energy. The quantitative survey (Palmgren et al., 2004) used wording taken from the interviews and reviewed by experts for accuracy and by nonexperts for comprehension. Respondents rated how much they favored or did not favor CCS and ranked nine low-carbon options. The ratings were below neutral and decreased after people were given detailed information. Compared with other technologies, CCS ranked far below renewables and even further below nuclear power. However, these results may not give an accurate picture, as the respondents were not given equally detailed information about the other options or about realistic portfolios combining various options.

In another study (Fleishman et al., 2010), the researchers developed materials about 10 technologies and 7 portfolios, again with input from experts and nonexperts. The research group produced a technology information sheet for each technology, discussing how it works, carbon emissions, other pollution, costs, reliability, and safety, and they prepared comparison sheets that allowed respondents to compare them on each dimension. They also provided seven portfolios of technologies, each of which would reduce carbon dioxide emissions by 70 percent compared with pulverized coal power plants. A total of 54 participants studied the material at home, were given more information in a session at their community organizations, and then held a group discussion. After this process, participants answered 86 percent of the factual knowledge (true-false) statements correctly. People felt they understood the issues, and the group discussion did not change attitudes much. People preferred coal plants with CCS over coal plants without it. Acceptance depended on the type of coal plants as well as on whether CCS was in the portfolio.

The next steps in the research will be to use national samples and to try a decision tool that will allow people to build their own portfolios. At this point, the results suggest that better information and risk communication may actually increase acceptance of CCS. A remaining question concerns what to do with people who will not study the material in detail. The group may try to put the information on the Internet with hypertext links.

In the discussion, Tuler noted that Judith Bradbury's study of CCS emphasized the importance of trust, for example, in the U.S. Department of Energy, as a determinant of acceptance. Bruine de Bruin noted that trust did not become an issue in her research, possibly because it did not discuss siting but only addressed the technology in general.

Pidgeon questioned why the initial reaction to CCS is negative, whereas the initial reaction to nanotechnology is positive, and he suggested that it might be because of negative attitudes to coal. Bruine de Bruine suggested that people see CCS as waste and want to get rid of the waste.

Anthony Leiserowitz asked about the views of the people who will be the opinion leaders on the CCS issue. Bruine de Bruin said that her group has shown that it can produce useful information materials, which would be useful for opinion leaders, most of whom are not CCS experts. Pidgeon noted that on the nanotechnology issue the initial positive reaction from Greenpeace in the United Kingdom influenced other environmental groups to join the process—something that did not happen with genetically modified foods.

Dietz commented that the advocacy coalition framework, which says that only policy leaders matter and that social learning in that community is critical to acceptance, has not made contact yet with the literature on risk communication.

COMMENT

Robert Marlay[5]
Climate Change Technology Program, U.S. Department of Energy

Robert Marlay said that the topic under discussion has been siting single facilities, but the scale actually needed is unfathomed by the general public. The world needs to avoid emissions of 2,000-3,000 gigatons of carbon (GtC), or 30 Gt/yr globally, or 5-6 Gt/yr for the United States. And 1 Gt/yr is what would be produced by 640,000 wind turbines or 1,200

[5]Presentation is available at [url] http://www7.nationalacademies.org/hdgc/Comment_Public_Acceptance_of_Energy_Technologies.pdf [accessed September 2010].

CCS coal plants, or biomass fuel production from a barren area 20 times the land area of Iowa. The journalist Thomas Friedman has commented that there is tremendous resistance to change in the United States because of polarized politics and debates and because corporations no longer act as citizens of the country. Friedman says the country needs better leaders and more courageous citizens.

The U.S. Department of Energy (DOE) has a bad legacy, but it has recently helped assist states with siting. It has learned that gimmicks, like hiring a movie star or an advertising firm, do not work. People want respectful dialogue at a realistic level of complexity. People tend to accept transmission lines, especially if they are convinced that there is a real need. To implement CCS, the process has to explore all the alternatives. Dialogue needs to be respectful, and planning and siting should be separated.

Marlay reported that DOE has a regional program on CCS with 21 regional partnership projects. The program addresses permitting, the regulatory framework, and public acceptance. It would be excellent to study this program. The process has 10 key steps. It cannot just be handed to a contractor, because it is important to have the right contractor. Unless this large process is handled right, there is no hope of solving the problem. Public acceptance issues must be addressed up front. He concluded by saying that there is no best practices manual for people who are trying to site facilities. He asked whether government should provide resources for public outreach.

COMMENT

Baruch Fischhoff
Carnegie Mellon University

Baruch Fischhoff noted that, in terms of strategic design, a little research goes a long way in giving useful insights. An insight from human factors research is that the need to communicate about a completed product or program reflects a management failure. It indicates a failure to incorporate the concerns of the relevant publics and to develop trust.

Strategic design needs (1) an appropriate philosophy (i.e., two-way communication); (2) leadership that promotes and implements the philosophy (he cited the Food and Drug Administration's Strategic Plan for Risk Communication approach as an example); (3) a staff that includes domain specialists, decision analysts, social scientists, and system specialists to make the engagement process sound; and (4) a methodology to deliver on the promise. There is a lot of science to advise on good design if the leadership is there to mobilize it.

Psychology research indicates that the public does sensible things if provided with good information and a sound communication process. As a result, communication failure suggests flawed management. Failures can arise from institutional inertia, inappropriate staffing (especially a lack of social scientists), isolation from lay people's concerns, indifference to lay concerns, and incentives for making the public look bad so decision makers gain power.

Social scientists can be part of the problem because of (1) a separation of analytical, descriptive, and intervention-oriented researchers; (2) isolation of researchers from practitioners; (3) a norm of making sweeping claims about people's competence as always high or always irrational; and (4) a predisposition toward manipulating rather than informing the public.

Finally, Fischhoff quoted the physical scientist Eric Barron's comment to the NRC Committee on Human Dimensions of Global Change, to the effect that social scientists do not know how to work together in order to do the big science that big problems need. That situation makes life easier for those who are predisposed to ignore social science evidence.

DISCUSSION

Kasperson expressed pessimism that the problem of social science capacity in government will ever be solved, particularly any time in the near future.

Marlay raised four questions or challenges for the group.

1. He said that DOE needs a best practice manual for handling public acceptance issues, guided by the National Research Council or by some of the participants in the room. He saw the need not for a cookbook but for a discussion of the options, identifying some successful models with their salient features. He said that DOE has few experiments to learn from.
2. He asked how it is possible to inspire better citizens without better leaders. By better citizens he means people who can engage in a process of dialogue that puts common interest above self-interest. He asked how the United States can become a can-do nation again. The environmental assessment process that began around 1970 is now highly evolved and largely successful, he said, asking if there is time to allow the things being done in energy and climate to go through a similar process, given the sense of urgency.
3. He also asked whether climate change communication is falling into the trap of the nuclear power debate, in which the experts say that anyone who disagrees with them is wrong. The climate deniers will come fully armed with arguments to take up time and delay action. He asked if there is "issue fatigue" on climate change, and if

so, what happens to acceptance of new energy technologies. Energy prices go up and down, but climate change is inexorable.

4. He noted Firestone's finding that, as more people were informed about offshore wind power, they became more favorable toward the technology, asking how that finding squares with public acceptance in Europe, which has gone in the reverse direction.

Tuler commented that there are cases in which siting has been quick and successful—the question is why they worked so fast. Fischhoff said that social scientists know how to make information comprehensible but the networks are not being created to inform people. Bruine de Bruin said that, in the case of CCS, the people who want to site the facilities are afraid to communicate with the public because they are afraid to draw attention to the project—so they put technical people in charge of the communication.

Several additional participants made comments or raised questions at the end of the session. Moser suggested that people tend to forget specific lessons of the past but retain general ones, such as the idea that "DOE cannot be trusted." Andrews said that although the discussion focuses a lot on how people can understand the technology, each community has a history, so decisions are path-dependent. Thus, communications need to start with the community rather than with fully developed proposals for what to site and where. Cohen noted that there are special issues about messaging, such as considering other messages already in place. For example, in western Canada, there is already active promotion of CCS by industry and government. New information needs to be customized to address the audience's preexisting knowledge and information. Rachelle Hollander noted that the issue of whether a community is being asked to contribute a fair share may also underlie some of the controversy.

4

Organizational Change and the Greening of Business

News stories indicate that some major corporations have undertaken serious efforts to reduce their carbon footprints. However, there has been very little systematic research into why they have done this while other corporations in the same industries apparently have not. The chapter begins with two leading scholars of organizational change presenting some general principles that might offer explanations. They are followed by an industry expert who discusses the results of an annual survey of business executives conducted by Johnson Controls, Inc., which reports on their attitudes about energy efficiency and the reasons they give for deciding whether or not to make investments in this arena. The following discussion, including contributions from people who have worked closely with business executives on climate change issues, covers other possible explanations for these business decisions. These contributions together offer a rich set of plausible hypotheses about what affects "greening" decisions of businesses. One discussant considers recent legislation aimed at changing business behavior and concludes by noting the need for social scientific study of business decision making to provide a basis for more effective legislation.

INTRODUCTORY COMMENTS

Andrew Hoffman
University of Michigan

Andrew Hoffman opened the session by emphasizing the importance of business organizations for supporting policy and for implementing solutions. Although policy debates have been dominated by discussions of the setting of a price for carbon, he said, a price for carbon alone will not achieve desired goals. Prices must be accompanied by institutional, social, and cultural change. For example, although business and the market responded well to the 2008 oil price spike, if the government had created the price spike, the response would not have been the same. In Ireland, a tax on plastic bags led to a 95 percent drop in their use. This happened not only because of the tax, but also through an associated change in norms, in which people who used plastic bags were considered loutish. Understanding how such a norm gets established is critical for understanding how markets will change to address climate change.

He introduced the session by saying that the first two presentations would cover social science knowledge at the individual and organizational levels. These would be followed by a presentation on what happens in corporations and then a discussant comment.

WHY DON'T WE ACT FASTER IN AN ENVIRONMENTALLY RESPONSIBLE MANNER? AN APPLICATION TO CLIMATE CHANGE

Max Bazerman
Harvard University

Citing his book, *Predictable Surprises* (Bazerman and Watkins, 2004), Max Bazerman reported that most people could identify problems in their organizations that were known to be worsening but that the organization would not address. Climate change is a great example of this phenomenon. People are very slow on the uptake. Why don't they act? There are cognitive, political, and organizational aspects, but his presentation focused mainly on cognitive biases, which are difficult to fix (Fischhoff, 1982). Several cognitive biases explain predictable surprises:

1. Positive illusions (it is not as bad as you think, there will be a technological fix, etc.).
2. Egocentrism: Who should fix it, those who created the problem or those who will suffer? People take more credit than they deserve when things are going well.
3. Overly discounting the future—people take benefits now and let others pay later.
4. Do not fix it if you cannot tell it is broken—but by the time you can tell it is broken, it is too late to act.
5. Bounded awareness—concentrating on some information keeps

people from seeing other things that are clearly there. The busier they are, the more they fail to see things that are plainly there.

Organizational barriers to action include (1) structural barriers (e.g., the problem of "silos"); (2) dysfunctional leadership incentives (e.g., no reward for incurring costs now for benefits to future leaders; thinking it's not my job to fix future problems now); and (3) failures of federal agencies to do their jobs, some of which can be traced to a failure to create significant campaign finance reform, which has led to money working against wise policies. Political barriers to action include the dysfunctional role that special interest groups play in preventing adoption of wise legislation.

In the discussion, a participant asked if there is danger of not having the big event that gets people to attend to climate change. Bazerman said that good advice needs to be available when a disaster comes, because that's when people will pay attention. He added that there may indeed be some big events.

Another participant asked why some businesses see it as in their interest to plan for a low-carbon future, and others do not. Bazerman agreed that there are differences and that leaders do matter. A participant pointed out that a lot of political action was taken on the environment in the 1960s and 1970s without a big crisis, suggesting that the rightward political shift in the nation has some responsibility. Another participant said that business leaders sometimes have "Aha!" moments. The example cited was Lord Brown at British Petroleum, who was said to have come to believe that climate change is happening when he was told that it is the reason he is losing his beachfront.

ORGANIZATIONS, INSTITUTIONS, AND IDEAS

Royston Greenwood[1]
University of Alberta

Royston Greenwood said that harnessing the power of organizations is needed to deal with climate change, but that they are very hard to change. He offered three main points.

First, organizations are socially embedded in an institutional context, which consists of the taken-for-granted ideas and values, often embedded in formal rules and policies that shape the way that people think and act. Because they are taken for granted, these ideas and values

[1]Presentation is available at [url] http://www7.nationalacademies.org/hdgc/Organizational%20and%20Institutional%20Barriers%20Royston%20Greenwood%20Univ%20of%20Alberta.pdf [accessed September 2010].

are regarded as normal and rarely explicitly considered. As an example, he reported that chief executive officers (CEOs) in the United States have more impact on the corporate bottom line than those in Germany or Japan, and he gave two explanations for this difference. One is a culture of individualism—leaders in the United States are expected to act as strong individuals and to make a difference. The other is an underlying institutional infrastructure that enables individualism, including distributed shareholders, boards of directors controlled by the CEO, and compensation arrangements focused on individuals.

Greenwood described the institutional context as having three components or pillars: (1) regulative, (2) normative, and (3) cognitive. Most attention in the climate change debate is being given to the first, the regulative. But such an approach underestimates the importance of normative and cognitive frames of reference (or "logics"). An example is the work of Rachel Carson (1962), who undermined the use of the synthetic pesticide DDT by changing how people think about its consequences and by undermining its normative acceptance. After huge resistance, she eventually changed national policy, but regulatory change followed change in cognitive and the normative context.

In fact, focusing on the regulatory context can produce unintended dysfunctions, such as (1) working to rule (a condition in which any specified rule becomes the maximum behavior), (2) rules being treated as nuisances with compliance becoming only ceremonial, and (3) government regulations becoming adversarial. In short, regulation has inherent limitations.

As an example of differences in normative contexts, Greenwood asserted that the "logics of action," that is, the underlying normative and cognitive ways that people think about how senior corporate managers should behave, have gone from a stewardship logic (in which the interests of multiple stakeholders are considered as relevant) to a shareholder service logic (in which only the interests of shareholders are considered). This latter logic is then underpinned by organizational structures and governance arrangements. An example is the linking of shareholder interests directly with those of the CEO through, for example, stock options. The corporate infrastructure gives the message that greed is good, and behavior changes accordingly. Changing this behavior implies a need to revert to the stewardship logic. In sum, the institutional context creates practices that are resistant to change. The normative and cognitive pillars are the most difficult to change, but they often receive less attention than the regulative.

Second, organizations are linked to the institutional context through organizational communities, such as industries, professions, occupations, and communities based on geographical proximity. These communities are important because they influence how corporations behave, especially when issues are unclear. The greater the ambiguity of an issue, the more an organization is influenced by its links to communities. Organizations look to their peers for guidance on how to understand and respond to issues. Change in the climate change debate may therefore depend on finding exemplars in an industry who can shape the "community" response either by example or by influencing the community through its formal associations. Looking at business communities may be informative when looking for ways to achieve change.

Third, Greenwood noted the potential importance of accounting firms as part of the institutional infrastructure. They link shareholders to corporations through the audit. An interesting way by which the climate change debate might gain traction would therefore be through the audit. For example, if regulators required that the carbon footprints of businesses be audited, such an audit would change the incentive systems in the businesses. There is evidence that the accounting profession in several countries is aware of this potential role and is developing appropriate training programs. The important point is that change in corporate behavior cannot be a function of formal government policy only, but should be reflected throughout the regulatory framework.

In the discussion, one participant pointed out that to get good environmental auditing, auditors need to be independent. Another participant noted the existence of huge corporate "spin machines" selling various messages with evocative language, such as "clean coal." Hoffman commented that the topic of organizational behavior is not popular in business schools, but it is very popular among business executives. Business school students think that they just need to come up with the right idea. Executives realize that the more important tasks are convincing people that it is the right thing and then getting them to do it.

BUSINESS ACTIONS ON ENERGY EFFICIENCY

Clay Nesler[2]
Johnson Controls, Inc.

Clay Nesler spoke about the Johnson Controls Energy Efficiency Indicator (EEI) survey, which includes responses from more than 1,400 executives responsible for managing, reviewing or monitoring energy use in their organizations.[3] Buildings are about 40 percent of the nation's carbon

[2]Presentation is available at http://www7.nationalacademies.org/hdgc/Survey%20Results%20on%20Barriers%20to%20Change%20in%20Business%20Clay%20Nesler%20John%20Controls%20Inc.pdf [accessed September 2010].

[3]Information on the survey is available at http://www.johnsoncontrols.com/publish/us/en/news.html?newsitem=http%3A%2F%2Fjohnsoncontrols.mediaroom.com%2Findex.php%3Fs%3D112%26amp%3Bcat%3D94 [accessed August 2010].

footprint and efficiency pays for itself, yet investment does not flow there. So the survey asked the executives about their attitudes, priorities, concerns, investment plans, and the decision criteria they used.

A total of 70 percent of respondents said that efficiency has never been more important, even in April 2009 when other economic issues were highly salient. This level of response was fairly consistent across commercial, industrial, and institutional buildings, although the highest level of attention was in the last category. When asked about new construction, 34-38 percent of executives said that it will be built to a green building standard (e.g., Leadership in Energy and Environmental Design), and about 17 percent said buildings would be retrofitted to a standard. On average, 60 percent said they add green elements to their design and retrofit projects but without seeking certification. The average organization expected to save about 6 percent from investments in energy efficiency.

Nesler noted that when Johnson Controls does energy performance analyses for retrofitting buildings, it usually estimates 15-25 percent savings. In their project to retrofit the Empire State Building, Johnson Controls expects to save 38 percent of energy, with a 3-year incremental payback.

Nesler also noted a long-term trend to invest less in the commercial sector. Managers of institutional buildings were planning to invest more, but the commercial sector lags for many well-known reasons. Respondents were expecting to see new legislation, and thought incentives in such legislation will be highly influential in their purchase decisions. And 57 percent of respondents said that climate change is a significant influence on their organization's energy efficiency decisions; 12 percent of the companies—mainly large ones—have public carbon reduction goals.

According to the respondents, the main barriers to investing in energy efficiency were lack of capital (40 percent), long payback (especially among industrial sector respondents, who expect a 2-year payback on average) (30 percent), and dedicated attention from ownership. The average "hurdle rate"—the level of expected payback or rate of return that an investment must pass to be approved—is 2.7 years, or a 35 percent rate of return. The hurdle rate was 2.9 years in the commercial sector and 4.3 years in the institutional sector, even though state laws allow investments with up to a 15-20-year payback. In short, although these investments are low risk with great and measurable rates of return, they are still not being made.

Nesler concluded by saying that there is increasing interest in energy-efficient investment among businesses, but investment has slowed. He expressed the hope that policy certainty, new incentives, and new financing solutions would lead to a surge in private-sector investments.

PANEL DISCUSSION

Andre de Fontaine of the Pew Center on Global Climate Change, who is trying to get companies to join a business leadership council on climate change, spoke about what he has observed in corporations. He reported that senior leadership is key to a company's involvement. Some companies have engaged recently in outreach to their entire workforce. For example, Alcoa has an online carbon calculator for employees to track their progress toward company goals, allowing them to compete if they so want. These goals are useful, especially if linked to action plans. United Technologies collects data on the carbon footprint of its facilities and uses the data in management. Toyota monitors and evaluates on the basis of British thermal unit, or BTU, used per vehicle produced. Companies are considering the cooperative benefits of efficiency, such as enhanced corporate reputation and better employee morale and attention to production processes. Some companies have set investment expense goals on an annual basis.

Melissa Lavinson of Pacific Gas and Electric (PG&E) identified a few drivers of corporate activity: the regulatory environment, the competitive environment, and the geographic environment. She said that customer and employee influences also matter. PG&E has evolved over 20 years, beginning with a chief executive officer (CEO) who created an environmental ethic and culture. The CEOs who followed were less aggressive on this matter, but the ethic was sustained through the people and procedures that had been put in place. The current CEO has brought in scientists to engage with the senior management team to discuss climate issues. He decided that the industry had a responsibility to develop solutions. He wanted to create a "line of sight" for employees, that is, a way for them to see how they could be part of the solution in their everyday jobs. The company also addressed the need to make the business case even for people who did not believe in climate change. The company developed a vision and value statement and links decisions with it, as well as new standard routines to make the new orientation last.

Nesler spoke more about the Johnson Controls project to renovate the Empire State Building The building today has hundreds of tiny offices, and the company that owns it wanted to attract larger and more up-scale tenants. The management company offered the building to the Clinton Climate Initiative (CCI), which said it needed an icon. The building owner wanted to prove that it makes good business sense to retrofit existing buildings and to create a replicable model for all large commercial buildings, so the Empire State Building was selected. All of the work will be in the public domain. The team that designed the retrofit program, which included Amory Lovins and the Rocky Mountain Institute, decided to remanufacture all 6,500 of the windows, adding mylar film "tuned" to the side of the building each window is on and filling the inner space with inert gas so that the windows will be super-insulated. The retrofit is expected to save 38 percent of energy, with a 3-year incremental payback. Johnson Controls, which is doing the retrofit, is guaranteeing the energy savings for 15 years. The owner says it would be irresponsible not to invest in this program. In sum, Nesler said that the real estate industry is starting to talk about its obligation to reduce emissions and the opportunity to save both energy and money.

Hoffman asked the panelists three questions. The first was about strategies for change. He remarked that although giving the CEO an epiphany is one strategy, there had to be more. What are the main obstacles? In his research, he found that accounting departments tended to be the most resistant to green investments. Several commenters noted the importance of procedures and incentive structures within a company. Lavinson agreed about accountants, noting that managers come and go, but accountants stay and weather changes of CEOs, so they need to be convinced or to have their incentives realigned. Nesler said that leadership is critical to get the process started, but not enough to keep it going. For example, some companies institute a corporate sustainability report, which looks good for a couple of years but then runs out of gas. He said that the goals have to be embedded in the culture of the organization. For example, Caterpillar created a "lean manufacturing" program, which meant that the environmental program put an emphasis on reducing waste. This put energy on the scorecard. The company developed measures of the carbon footprint of every project. De Fontaine added that when Walmart started rating its suppliers on their carbon footprints, that created a lot of change.

Hoffman then asked what happens when one company diverges from the rest of its industry. Lavinson said that in the electricity industry, a caucus of CEOs challenged the leadership of EEI, with some threatening to leave the institute. EEI's national position on climate change went against the interests of some of the members, eventually leading to changes in EEI. The industry has an aging infrastructure, so it needs to make investments for a 30-year time frame. Executives are concerned that if the investments are bad, there will be trouble from their customers. Coal-heavy electric utilities are beginning to accept the view that change is needed.

Nesler said that his company has invited its suppliers to disclose their greenhouse gas emissions publicly. There were very diverse responses, including a few suppliers who refused to respond at all. The company helped its suppliers learn how to do it. He said that Johnson Controls has seen that the things it does for environment are also good for its business and that other companies that track their emissions and manage their energy for a few years also learn that there are benefits both for the business and the environment. De Fontaine noted, however, that some companies, for example in the agricultural sector, get pressure when they get ahead of their customers.

Finally, Hoffman asked how companies' responses might change if suddenly there were a cap-and-trade system, or a carbon tax, or a renewable portfolio standard. Lavinson replied that a renewable portfolio standard would tell her company how to reduce carbon, whereas cap and trade would let the company decide from a broader portfolio of options. A carbon tax would act like cap and trade in this respect. Nesler said that Johnson Controls already factors in the future cost of carbon in its decision making. The company thinks more is needed than cap and trade or a carbon tax. There also need to be stronger building codes, appliance standards, incentives for retrofits, and renewable portfolio standards to get immediate emissions reductions. He said that the efficiency market is inelastic—there would be little impact on the demand for energy efficiency for any carbon price under $40 per ton. For that reason, a set of complementary policies is also needed.

In the subsequent discussion, Charles Wilson asked about ways to accelerate the diffusion of innovation into business communities. Will stories do it? Are formalized mechanisms using peer groups necessary? Nesler said that Walmart has a Personal Sustainability Program that gets groups of employees to work together on activities that promote general well-being (not only about the environment). The idea is that taking action personally will help it spread. Following that idea, Johnson Controls, which has a "vision week" every year, devoted it to sustainability 2 years ago. The company identified 30 things to do at home, at work, and on the road, to improve sustainability and asked employees to do one thing in each category. The company challenged them by setting different businesses and regions against each other. The company had 100,000 people involved, and they continued after Vision Week ended. An online version of this personal sustainability challenge is available in the public domain at http://www.mygreenprint.org.

Lavinson said that to get people to improve energy efficiency, codes and standards are by far the most effective. With some utility companies, it is hard to make significant inroads because of their business model. If a utility has a business model that is agnostic about saving energy versus selling it, it becomes an ally in addressing climate change. They have all the data they need. De Fontaine suggested that the most effective strategy is to make behavioral change automatic—to set lights to go off automatically, prevent thermostats from being set too high, and so forth.

Susanne Moser, noting that dialogue between social scientists and business leaders happens even less often than between social scientists and policy makers, asked the panelists what they would ask from social science. Nesler replied that business makes up the social science as it goes along.

He noted that it would not design a new product without an engineer, yet it launched a worldwide employee program without consulting a behavioral scientist. He went on to cite several programs in businesses, at the state level, and in nongovernmental organizations—all invented in similar ways. He noted that companies go to the American Council for an Energy-Efficient Economy to share ideas. The group has stakeholders and a budget, and he believed it could be a great experimental ground and would be very open to social science involvement.

Hoffman remarked that there are no incentives for social scientists to work with business. Social scientists have to translate their knowledge and make it accessible to business, but the incentives are aligned against that. Bazerman suggested that people with good ideas might be able to contact Nesler or Lavinson with a one-page précis. But he went on to note that for social scientists to publish, they need to run experiments. As a social scientist, he would want to suggest that a business try an idea on half its employees to produce a high-quality study. It would be possible to mix a high-quality experiment with a good demonstration project.

WHAT YOU DON'T KNOW ABOUT HOW PROPOSED FEDERAL CLIMATE CHANGE LEGISLATION WOULD HARNESS CORPORATE AND INDIVIDUAL BEHAVIOR

John Dernbach[4]
Widener University Law School

John Dernbach began by saying that law is all about behavior. However, those who draft legislation guess about how people and organizations will respond to what they are doing. For example, he drafted the Pennsylvania mandatory recycling law in 1984. It requires curbside recycling because practitioners all said to make it easy to comply.

He said that the movement to behavioral economics can be seen in proposed federal legislation. In his view, the federal climate legislation creates a norm in addition to the regulations. He sees utility companies as a major obstacle because of state laws that encourage the consumption of energy. Companies that lead on climate change have multiple influences—what is good business, what is good for environment, what Walmart is doing, and so forth. These companies will have a competitive advantage, which will help reinforce the norm when the advantage is seen.

Regarding federal climate legislation, Dernbach said that all the major bills have cap and trade as a centerpiece. This will raise the price of carbon, which will motivate individuals to efficiency and conservation, although

[4]Presentation is available at http://www7.nationalacademies.org/hdgc/Comment_Public_Acceptance_of_Energy_Technologies.pdfPg. 4-8 [accessed September 2010].

measures to reduce the price impact on consumers would reduce this effect. He emphasized that market imperfections will limit the effectiveness of the price signal because (a) consumers consistently undervalue efficiency and conservation savings, (b) there is a significant principal agent problem (e.g., in rental housing), and (c) the price signal is not sufficient to trigger the investment needed to get an 80 percent reduction in emissions. Also, as has been made clear in many presentations at the workshop, prices alone are insufficient to motivate individuals.

Dernbach went on to note that these bills have a number of other provisions to deal with the market imperfections. One of the key behavioral provisions is a State Energy and Environment Development Account in H.R. 2454 that would fund energy efficiency programs, such as building codes, building energy performance labeling programs, and low-income community energy efficiency programs. The bill would also create a new Clean Energy Deployment Administration in the U.S. Department of Energy (DOE) that would publish goals for the deployment of such technologies as a zero net energy building stock and make financial products available to enable owners to make energy efficiency investments with reasonable payback periods. It would allow DOE to offer bonuses to retailers or distributors for sales of best-in-class appliances; bounties for replacement of old, inefficient products; and reward manufacturers for producing new, super-efficient products. It would also support communication and rebate programs for water-efficient products. Both H.R. 2454 and S.B. 1733 would also address transportation planning, including encouragement of public and nonmotorized travel, zoning changes, and travel demand management programs. Both bills would call on the U.S. Environmental Protection Agency (EPA) to study and then develop a product carbon disclosure program and report back on the results. They would also provide credits for reforestation and other activities to offset carbon emissions.

Dernbach concluded that to make these proposals effective will require social science. What is missing in the legislation is an effort to fully engage the public. There is a need to use what is known about motivating human behavior to engage people nationwide. Environmental law has traditionally treated polluters as "other," and this legislation follows that model. What is needed now is to engage everyone.

In a brief concluding discussion, Baruch Fischhoff noted that the Food and Drug Administration and the National Cancer Institute both know they need behavioral science and have some staff on hand to make changes. EPA's Science Advisory Board, however, chose not to protect its behavioral science and wants to restore its 1970 worker profile. EPA does not, in his opinion, even know who to hire to get the behavioral expertise it needs. The organization is not structured to do this. Dernbach replied that ideally the legislation would call for engagement and agency officials would follow that lead.

Part II

Adapting to Climate Change

The second workshop, held on April 8-9, 2010, convened researchers, practitioners, and federal officials to bring available knowledge and experience to bear on a set of questions that are likely to be critical in shaping a national strategy for adaptation to climate change (see Box II-1). The questions were developed by the organizing panel on the basis of input from practitioners responsible for developing adaptation plans. They were distributed to all participants in advance of the workshop. Presenters were asked to consider the questions in making their presentations, but they were not required to address each of them explicitly. Participants were advised to think about key concepts and knowledge relevant to climate change adaptation, focusing the knowledge on such practical questions as those on the list.

Much of the knowledge on adaptation to climate change derives from case studies in particular places or focused on managing particular types of resources that may be affected by climate change. This knowledge is not yet organized around a generally accepted theory or a unifying set of concepts. However, there are concepts and bodies of knowledge from other fields that might help add coherence in this field. In order to contribute to an increased coherence of knowledge, the panel began the workshop with presentations and discussion about the state of the science, including the state of theory and concepts, and it asked selected panel members to offer, at the end of the workshop, their syntheses of the possible answers to the practitioners' questions that energed from the presentations and discussions. Chapter 10 presents these panel members' syntheses.

BOX II-1
Key Questions About Climate Change Adaptation

Initiating adaptation efforts:

1. What are the main barriers to initiating adaptation efforts, and what has been effective at overcoming them?
2. How and under what conditions have climate change considerations been successfully integrated into the normal activities of regional or sectoral risk management organizations?

Coordinating adaptation efforts:

3. What strategies or methods have been effective for coordinating adaptation efforts across scales (e.g., national, regional, local, individual)?
4. What strategies or methods have been effective for coordinating adaptation efforts across sectors (e.g., government, private, nonprofit, community)?
5. How should stakeholders and the public be engaged in adaptation efforts?

Informing adaptation efforts:

6. What methods have been successful in providing needed information to risk managers who must cope with climate change?
7. How should efforts to inform climate adaptation characterize risk and uncertainty about future climate and other processes affecting climate risk?

Science needs for adaptation efforts:

8. What new social science knowledge is needed to develop a national adaptation strategy?
9. What metrics and indicators are needed to support adaptation decisions (e.g., indicators of vulnerability, resilience, adaptive potential, effectiveness of adaptation efforts)?
10. What are the key needs for databases, scenario development, and modeling?

Managing adaptation efforts:

11. How should a national climate adaptation effort set priorities across hazards, sectors, regions, and time? What criteria, and what processes, should be used?
12. What mechanisms can help make adaptation efforts more adaptive? How can a system enable decision makers to learn efficiently from experience?
13. What additional capacity do federal agencies need to support adaptation and resilience?

5

Climate Change Adaptation: The State of the Science

INTRODUCTORY COMMENTS: WORKSHOP OBJECTIVES, CONCEPTS, AND DEFINITIONS

Roger Kasperson

Roger Kasperson, the panel chair, opened the workshop by noting that its goal is not to develop a research agenda, but rather to identify some areas in which the social and behavioral sciences already know enough to be helpful in developing societal responses to climate change. He noted that the workshop organizers had circulated a series of critical questions they would like the workshop to contribute to answering (see Box II-1). He described the planned organization of the workshop. First, it will assess the current state of knowledge. Second, it will discuss some of the efforts to address the issues through federal policy. Third, it will examine a series of case studies in several panel discussions. Last, the workshop participants will return to the initial questions and consider whether the community has any answers. Kasperson stated that he might be the most skeptical person about whether the adaptation community knows as much as it needs to know to assist societal efforts in this domain. He hopes the workshop can separate what is known from what participants would like to know.

With respect to the relationship between science and decision making, Kasperson noted that there are two main metaphors. One is of a bridge, a pipeline, or a superhighway between science and practice. However, research and experience suggest that what exists looks more like a spider web, with multiple centers that move around. In Washington, people talk

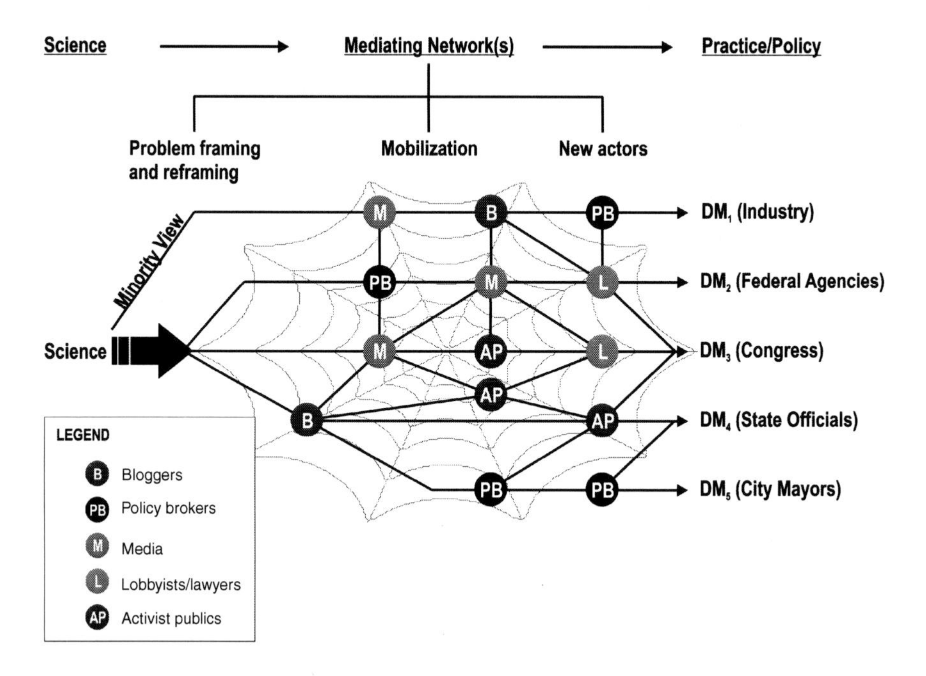

FIGURE 5-1 A hypothetical web representing the relationships of science and decision making.
NOTE: DM = decision making.
SOURCE: Roger Kasperson. Used with permission.

repeatedly about a linear process that starts with the science and then applies values at the end of a management process (if any funds still remain). Kasperson said that the process is actually much messier. He presented a hypothetical schematic to represent the metaphor of a web (see Figure 5-1). Webs may be expected to take very different forms for different cases. Insights and ideas come out of science and go through a process of mediation by many diverse actors, with some of them disappearing and others being elevated in importance before decision makers ultimately act.

There are simple webs, with strong linkages of science and decision making through what have been called boundary organizations. Some webs are complex and stable; others are dynamic and unstable, with actors appearing and disappearing. A complex and unstable web is probably descriptive of climate policy—and there is limited understanding of the shape or functioning of such webs.

ADDRESSING STRATEGIC AND INTEGRATION CHALLENGES OF CLIMATE CHANGE ADAPTATION

Ian Burton (with the assistance of Thea Dickinson)
Meteorological Service of Canada and University of Toronto

Ian Burton began by observing that the topic of adaptation to everyday climate has been around for a long time under different guises. Adaptation to anthropogenic climate change is another matter. The short version of the story begins with the United Nations Framework Convention on Climate Change, signed in 1992, which put adaptation to climate change onto the public agenda. Because that convention was focused on greenhouse gas emissions, it emphasized pollution control and mitigation, following the model used for addressing ozone depletion in the Montreal protocol. Thus, the scientists who advised politicians focused attention on pollution control. However, the developing countries that considered themselves most at risk and least responsible insisted that adaptation be written into the convention, and rich nations agreed to contribute to the costs. The problem was defined as adaptation to climate change, as opposed to adaptation to climate, and this distinction has hung up discussion ever since. The most recent Intergovernmental Panel on Climate Change (IPCC) report (Parry et al., 2007) defined adaptation in terms of "adjustment."

In this short story, adaptation to climate change was born in 1992 and was initially of concern to developing countries, which hoped for additional development assistance as a result. That perspective led to a lot of research on adaptation to climate change. The knowledge base can be defined broadly to include all of social science knowledge on the subject. There is a much longer tradition of research on adaptation to climate change in the global South than on adaptation in the rich countries.

Policy developments on adaptation since 1992 include these milestones. The Kyoto protocol, adopted in 1997, imposed a 2 percent levy on agreements under the clean development mechanism to support adaptation. The Marrakesh accords of 2001 called for national adaptation plans of action in the 49 least developed countries. By the 2007 Bali conference, four "pillars" of climate response were identified: (1) mitigation, (2) adaptation, (3) technology transfer, and (4) finance. At the Copenhagen meeting in 2009, there was a promise of substantial additional funding for adaptation and mitigation in developing countries, up to $3 billion (U.S.) by 2012 and $100 billion per annum by 2020. One might reasonably ask whether there is any real chance that these promises will be kept.

Meanwhile, there has been little progress on mitigation. Mitigation is now recognized as a problem of emerging economies as well as in the global North. There also has been a belated realization that developed countries share the risks of climate change. The earlier idea that mitigation is global but adaptation is local has collapsed. It is now widely recognized that adaptation needs national and international cooperation to succeed. Climate change also has been recognized as an issue of development and equity, not only a pollution issue.

Synthesizing a statement of the knowledge base is certainly part of the way forward. But there is also a longer version of the story. Moving forward can start by looking back. The knowledge base did not start in 1992: when the idea of climate change adaptation became a focus of policy attention in 1992, the existing disaster risk management community was astonished and said it already had a lot to say. Much of the research in this area has been reviewed by Dennis Mileti in *Disasters by Design* (Mileti, 1999). As far back as 1945, Gilbert White said that while floods are acts of God, flood losses are the results of human choice. There was a great deal of knowledge about adaptation to climate before 1992, but most of it was based on the assumption of a stationary climate. There has been a systemic failure to deal with climate extremes as well as society could and should, even before climate change came into the picture.

Climate variability and extremes had been considered in terms of events, from which social systems recover and return to normal; now sequences of events are considered, such as progressive series of droughts or floods or cyclones, as well as cumulative desiccation and sea level rise, rather than just isolated droughts and storms. This change in view has led to more of a focus on systemic risks and to thinking in longer terms about risk reduction rather than only about assistance for recovery. Disasters once were considered as humanitarian concerns to be dealt with one at a time; now there is an emerging idea that the recurrence of disasters is expectable and that they are a common responsibility. The two communities are now coming together.

There is a deficit in attention to adaptation, even in wealthy countries. Despite the expansion of physical science knowledge, disaster losses are increasing globally. Efforts at natural hazards management, human adjustment, disaster risk reduction, and climate change adaptation have not been successful. Can knowledge be better marshaled to address the adaptation deficit?

Burton suggested talking about forensic disaster studies, asking,Who or what is responsible for disaster losses? Responsibility is widely dispersed. Rich people choose amenities over risk reduction, the poor often have little choice, and government assistance to victims of disasters can create moral hazards. What is known about flood insurance, and is there a credible assessment of how it works? A research proposal to look at this topic is being sent to the International Congress of Scientific Unions and to national research councils. The research would look at climate risks the same way that transportation analyses look at the causes of traffic accidents, to pin down responsibility. Perhaps this kind of analysis can be done for stationary climate, and then climate change can be added.

Burton ended by saying that there are many challenges. Adaptation is local, but also regional, national, and global. It is multisectoral, so all the sectors must be part of any national adaptation strategy. An interagency task force may not be sufficient to address the problem, and a more integrated approach may be required. He reminded the audience not to forget mitigation. Adaptation choices may have important near-term benefits by reducing greenhouse gas emissions and generating other desired outcomes. They may also have adverse long-run consequences, including increased emissions.

ADDRESSING BARRIERS AND SOCIAL CHALLENGES OF CLIMATE CHANGE ADAPTATION

Neil Adger[1]
University of East Anglia

Neil Adger addressed three issues: (1) the roles of the multiple actors that are adapting, from civil society to markets and government; (2) equity issues related to who is vulnerable and who makes adaptation decisions; and (3) barriers and limitations to implementation of adaptation measures. In looking at the future, people not only use scenarios and models but also observe adaptation and make inferences from past adaptations to variability in weather and climate. They can also learn from observing ongoing adaptations to anticipated climate change.

The United Kingdom (UK) provides some good examples. There are efforts to build adaptive capacity, to regulate land and water for future use, and to implement some adaptive actions (e.g., coastal defense, as in the protecting the Thames estuary against anticipated risks to 2100). Actuaries are calculating insurance premiums for the future. There are also efforts to provide public good information to stakeholders through the UK Climate Impacts Programme. The UK Department of Health is doing planning for heat waves.

In 2005, the Tyndall Centre identified 300 examples of adaptation—mostly in governments at different levels, but many in the private sector as well. Government in the United Kingdom is in the vanguard of adaptation. If there are best practices in government or the private sector, however, it is

[1]The presentation is available at http://www7.nationalacademies.org/hdgc/Addressing%20 Barriers%20and%20Social%20Challenges%20of%20Climate%20Change%20Adaption.pdf [accessed September 2010].

not yet clear if they get diffused. In fact, so far there are no assessment criteria for judging whether these adaptation efforts were effective. Research on diffusion of technology may offer models that can show how widespread the diffusion of adaptation will be. The issues affecting implementation of adaptation are cost, timing, power, responsibility, equity, and irreversibility of impacts (see Table 5-1). An obvious question is whether the first focus should be on trying to diffuse practices or on making sure they are effective. Generally, anticipation is believed to cost less than adaptation after the fact.

Adger noted that resilience and vulnerability are not antonyms. He distinguished three normative goals or principles for government intervention: protecting the most vulnerable (Rawls), efficient adaptation (Pareto optimality), and system resilience (rather than a focus on individual parts). He pointed out that if an ecological system is moving from one state to another, resilience could have a number of different end states. The idea of measuring vulnerability presumes a threshold of risk beyond which a population is vulnerable. There may be parts of a population that are vul-

TABLE 5-1 Decision-Making Questions

Issue	Key Question	Outcome of Indecision
Cost	Who bears the costs of adaptation?	Costs may be shared unevenly in terms of willingness or ability to pay
Timing	When is action taken?	Anticipatory adaptation may be cheaper: ex post recovery has greater social costs
Power	Who makes the decisions?	Inaction or stalemate between advocates of national and local views
Responsibility	Who takes action?	Responsibility inevitably devolved to individuals
Equity	What kinds of change are acceptable?	Unacceptable risks are imposed on the least powerful people and on public or private infrastructure
Irreversibility	How are irreversible impacts considered?	Increased economic cost of future options lost; asymmetry in loss aversion

SOURCE: Amended from Tompkins et al. (2008, Table 2). Used with permssion.

nerable and a proportion that is not. If so, it makes sense to concentrate on the vulnerable proportion of the population. It is possible to measure vulnerability by the proportion of a population that is vulnerable or by the distance of people from the threshold.

Social science has a lot to say about vulnerability, but Adger questioned whether there will be the luxury of time. Fairly radical adaptation is likely to be necessary. He noted a new realism about the need to plan for adaptation to a greatly changed climate. He also noted that adaptation may be limited because people have diverse and incommensurable values, because foresight is uncertain, and because people place intrinsic value on current places and identities. Adaptation is also constrained by social characteristics, individual behavior, and other barriers.

Adger spoke about the potential and limitations of markets for responding to climate change. Critical adaptation needs concern water resources, property loss, human health, nature conservation, and cultural heritage. Markets might solve problems at the top of this list, where they work fairly well, but less so further down in the list. Citing work by Farber (2007), Adger said that markets work better for "geographic" impacts—water, coasts, habitats—but not well for diffuse impacts (e.g., on global food systems) or for catastrophic climate changes at the global level. The issues that limit markets are those that involve social externalities (effects on communities or places) and loss of nonmaterial assets.

The perceptions of the vulnerable affect the ability to adapt. Research by Adger and colleagues (Abrahamson et al., 2009; Wolf et al., 2010) has found that among elderly people in the United Kingdom who are vulnerable to heat waves, low self-efficacy reduces action to adapt. For elderly people living alone, their ability to live independently was important to them, and many denied their vulnerability by denying that they are elderly. He pointed out that heat wave planning will not work if targeted to the vulnerable, if they deny that they are vulnerable.

Adger noted as well that many communities will resist planning for climate change and instead will actively lobby for protection against it. Some UK coastal communities, for example, actively resist sensible plans for adaptation. What matters to them is control over the process of planning their community's response. Well-functioning property markets may create incentives to adapt by devaluing vulnerable properties, but adaptation still will not be easy because of community identity and other issues.

Adger identified three emerging issues: (1) Can people learn about planning to adapt from coping with crises? (2) Individual perceptions of resilience and vulnerability are drivers of social processes of adaptation. (3) Social inertia in the form of strong attachments to the past or to current conditions can be a significant barrier to adaptation.

DISCUSSION

In comments at the end of the session, Anthony Janetos mentioned a Heinz Center report on what is known about flood insurance. He said that the U.S. national flood insurance program is nothing like what the industry would create; it is more of an income distribution program. Helen Ingram asked about how to shift the focus from particular disasters to the larger, systemic picture. Ashwini Chhatre asked about current maladaptations that need to be undone, such as property development on the Florida coast and irrigation in California's Central Valley. He also asked how to address silent disasters that are not in the headlines. Roberto Sanchez-Rodriguez expressed surprise at the lack of mention of the structure of governance. He asked what provides stability in the webs Kasperson mentioned.

In response, Burton commented on insurance internationally, noting that there has been talk of an international public-private partnership, and that there are experiments with new insurance products in developing countries. Insurance could be used not only to spread risks but also to create incentives for adaptation. He said that people need to learn from handling extreme events and to build this learning into planning for "creeping" disasters. Burton said that disasters are part of everyday life, and that what makes disasters serious is embedded in society. People need therefore to look at them in social terms, identifying how to reduce them in a precautionary manner, cope with them when they occur, respond to them after they happen, and absorb and use lessons for future events.

Adger responded by underlining Janetos's point that government interventions may not help improve adaptation. Societies are bound by demands to maintain the status quo. For example, drought recovery programs in Australia took a lot of blame for damage from fires because the government was subsidizing farmers to stay in dry areas with high fire risk. Australians like their rural populations and will pay to support them even though it is maladaptive.

Adger said that the United Kingdom is trying to get away from learning from disasters by being more proactive. The UK government is very proud of its anticipatory planning for the Thames for 2100, but it is hard to bring the population along. Planning agencies are planning for 2 million new houses, many of them in vulnerable areas in the Thames gateway. Adger does not think there is an optimal governance approach for adaptation. In the United Kingdom, for example, local government structures face different cultural identity issues. In the Orkney Islands, people looked at climate change as an opportunity for more local control; in other local communities, the dominant view was that the problems should be addressed nationally.

Kasperson noted that systems ecologists have pointed out that risk management interventions often serve to keep a system from moving to a new and more resilient state. He asked how policy can get out of that quandary. Maria Carmen Lemos pointed out that to address this, one needs to know the desirability of the state one is moving into—and people either do not know what their collective preferences are, or they do not agree. Thomas Dietz noted that people want to preserve their sense of place but have a short time horizon. They try to preserve what they personally remember, not an ancient past. He added that there is not much knowledge about this problem.

Maria Blair said that people fail to recognize the concept of social inertia. The whole concept of adaptation suggests change, which people often oppose. She suggested that allowing for processes with local control or engagement might address this problem. Adger noted that his examples of social inertia focus on things people do not want to change. On the mitigation side, however, there may be support for changes if they are seen as ways to improve welfare. For example, the UK "transition towns" movement tries to make towns more resilient in a future after "peak oil" by increasing local food production. However, many people affected by weather-related risks see a relation to climate change, yet they are not thereby motivated to reduce their emissions.

Richard Andrews pointed out that major change usually happens during windows of opportunity. Posthurricane planning, for example, can provide such a policy window. Planning can be based on the recognition that a community cannot afford to rebuild just the way it was and has to become more adaptive. It is important to be consciously ready in advance to move when those moments arise. Burton noted that in postdisaster situations, however, there is always tremendous pressure to return to the status quo. Andrews said that Florida, for example, has been maladaptive in having the state become the primary insurer of coastal property when market rates became politically unacceptable, whereas in North Carolina, state policy has struck a less morally hazardous balance that combines increased coastal insurance rates with a safety valve to assess all state property owners for shares of losses above a high threshold of economic catastrophe. Burton suggested that mitigation allows many more options at an individual level, whereas most adaptation requires collective action, at least at the community level.

Christopher Farley questioned the value of terms like "maladaptation." The physical systems are extremely complex, and society has put policies in place. He said that during the 1990s timber wars, the U.S. Forest Service could have the "right" answer from the science, but society could insist on action another way. He said it is important to create systems and processes that allow society to make decisions with awareness of what the impacts are. He also suggested that there is no way to draw the line between who is vulnerable and who is not—in fact, the insistence of many elderly people on *not* defining themselves as "vulnerable" is very positive and adaptive in many ways—and that negotiation might be needed to draw such a line.

Jamie Kruse noted that the time scale of adaptation has to match the changing conditions. "Social inertia" is a mismatch between what a group, such as the workshop participants, believes to be the right choices and what society is doing.

6

Federal Climate Change Adaptation Planning

THE INTERAGENCY CLIMATE CHANGE ADAPTATION TASK FORCE

Maria Blair
White House Council on Environmental Quality

Maria Blair began by saying that the Interagency Climate Change Adaptation Task Force was established in early 2009 under the leadership of the White House Council on Environmental Quality (CEQ), the Office of Science and Technology Policy (OSTP), and the National Oceanic and Atmospheric Administration (NOAA). Some 23 federal agencies and offices and close to 350 people are involved in the task force's work. The initial mandate was to make recommendations on adapting to climate change, both nationally and internationally. The activity responded to three questions: (1) How does the federal government deal with adaptation in its own programs and operations? (2) How does the federal government best support adaptation activities at lower levels of government? (3) How should the United States help other countries with greater vulnerability build resilience, especially given considerations of the effects of climate change on national and homeland security, development assistance, and, through market effects, on global supply chains?

In October 2009, President Obama signed an executive order focused on greenhouse gas emissions reduction in the federal government, which included a section that recognized the task force and called on it to develop "recommendations toward" a national adaptation strategy by October

2010. The task force has not been asked to produce a national adaptation strategy by that time. Blair indicated that, by then, the task force would not even be able to produce a suite of recommendations for the issues currently defined, let alone future issues. She anticipated that the October 2010 report would offer some substantive and process recommendations for federal government action.

Blair stated that the interim progress report released on March 16, 2010, does not contain many recommendations but states three important conclusions to which the 23 agencies agreed.[1] First, climate change risk and adaptation opportunities are critical issues for the United States. Second, the federal government must adapt and improve resilience. Third, the task force has begun working to understand the implications of climate change for its work domestically and internationally.

The task force found that there is substantial activity under way already in the federal government and in the country. Some U.S. states, cities, and counties have begun to assess risks and opportunities and to adapt and build resilience, as have other countries. The federal government is taking action through several different agencies; however, there are significant gaps. These include the lack of a unified strategic vision; of an understanding of the challenges at all levels of government; of organized and coordinated efforts across scales; of strong links between support and participation of tribal, regional, and state governments; of coherent research programs to identify and address impacts and of relevant and accessible impact information for decision makers; of comprehensive and localized vulnerability assessments; of budgetary and other resources; and of a robust approach to integrating these issues and learning and applying lessons. So although the government is engaged, it still lacks many of the needed building blocks.

The interim progress report highlights two themes. One is mainstreaming or integrating: climate change adaptation needs to be part of the everyday decisions and core missions of all affected agencies, rather than being the job of a separate adaptation office. The other is the need for a forward-looking, flexible approach. The past should not be the sole input to decision making: some agencies will need to make significant changes in how they make decisions.

The report lays out a minimum set of components of a national adaptation strategy. The first is integration of science throughout decision making and policy, from the physical to the social sciences. A second is communication and capacity building: there is a need to develop a good way to talk about adaptation that engages people and invites them into a decision-making process, and there are real challenges to the ability to

[1] See http://www.whitehouse.gov/sites/default/files/microsites/ceq/20100315-interagency-adaptation-progress-report.pdf [accessed September 2010].

understand key needs for capacity building, even within the federal government. The third requisite is coordination and collaboration, both of which are critical in the development of new approaches. Fourth, prioritization criteria are needed among the actions to be taken, especially for interagency and government-wide action. Fifth, a flexible framework is needed for agencies to use in integrating climate risks and adaptation measures into their missions and operations: the task force is piloting a set of principles for agencies to use. Sixth, evaluation (learning) is essential, and the process of addressing climate adaptation must itself be adaptive over time and experience: people need to learn from the actions they are taking and to change as they move forward.

The task force's next step is to report by October with initial recommendations, including some near-term and some longer term process recommendations. It held a set of listening sessions during fall 2009, another set is going on now, and a series of regional outreach sessions is planned both nationally and internationally.

Blair offered some additional observations related to the questions of concern at the workshop. The adaptation planning process was started in the government by two scientists (Jane Lubchenco and John Holdren) and by Nancy Sutley, CEQ chair, who had previously worked on water issues. The need for integration of adaptation into many agencies' missions and operations and into their cooperation with each other is a core principle for advancing effective adaptation. Although the task force is seeing momentum and interest in some places, in others social inertia prevents any movement at all. The lack of capacity is a major barrier to further progress: there is a need to invest in the skill set needed for adaptation across many agencies. In Blair's view, there is a tension between mainstreaming adaptation into all the relevant agencies' operations and building the distinctive capacity for addressing adaptation that is also needed. The task force does not propose to create a new "adaptation office" in the federal government, yet without some source of concentrated expertise on it, there is a real gap in capacity. The government also lacks effective approaches for internal coordination and collaboration or a good bridge for collaboration with state and local governments and the broader public.

The task force process has also focused more on how to use what is understood today than on defining an agenda for further research that is needed. There are some areas of research coming to their attention (through disaster researchers, for example), but these are not yet systematically or carefully selected to include all relevant bodies of knowledge. In fact, not many scientists have been involved with the task force: those involved have been mainly practitioners. Also, there are other places in the government to address the research issues (such as the U.S. Global Change Research Program and the National Assessment of the Consequences of Climate

Change). In conclusion, Blair said that the task force is not yet ready for management, as it does not yet have anything to manage. She added that flexibility is a real challenge for the federal government.

NATIONAL ASSESSMENT OF CLIMATE CHANGE

Kathy Jacobs
Office of Science and Technology Policy

Kathy Jacobs spoke briefly about the new National Assessment of Climate Change now being organized. She said a major item of discussion is how the federal government will change the connection between science and decision making in this assessment. In the past, the national assessment process was focused on writing a report on the state of vulnerability. The desire now is to focus on a process rather than a report—a process to reduce vulnerability across the country. The assessment would be framed in terms of decision support rather than providing a summary picture. She said the nation has not benefited as much as it should because the national assessment was organized previously to meet regulatory goals. Now the government wants to use the assessment to build national capacity. She said that there is no dedicated budget for the national assessment and that only NOAA has even requested any funds for it. The leaders of the process are trying to build a process that communities own, as well as a long-term system for evaluating vulnerability and risk across the country by knitting existing observational systems together. The intent is to use the process to influence the focus of federal investments.

DISCUSSION

Susanne Moser asked Blair how resilience is understood in the assessment group, and also what it is being mainstreamed into. Blair said she began by hating the word "mainstreaming" because she thought it was an excuse for ignoring the issue. But she now thinks that the challenge of changing the decision system is even larger than the challenge of adaptation. She said she would like for the federal government be at the point at which the Army Corps of Engineers and the U.S. Department of Transportation take climate projections into account in making their infrastructure decisions—even though the system for doing so is very imperfect. She added that there is no federal conversation about radical change in the decision-making systems.

Carolyn Olson said that an audience at Agriculture Canada liked the task force's definitions for adaptation and resilience.[2] She also suggested that the adaptation issues are structured differently in different agencies, noting that U.S. Department of Agriculture is a department that traditionally links research to extension and is thus different from some other departments.

Stewart Cohen asked if capacity building includes creating more extension agents. He questioned, for example, if engineering training should include training in climate change, and whether such cross-training could make extension efforts more effective. Blair replied that the idea of extension as part of capacity building is interesting. She said that thinking about how different agencies will approach that will be an ongoing process.

Rick Piltz asked about coordination needs among federal agencies, and federal-state-local coordination. He saw an obvious need for a national adaptation preparedness office and could not understand why the federal government is not considering this.

Blair said that the task force is focusing on coordination challenges among agencies and on federal-nonfederal coordination, not on intra-agency coordination. She said that the possibility of a central office is not off the table, but the task force wants to make sure that everyone pays attention to climate adaptation and does not delegate the issue to specialists. The task force is trying to learn from successful models, but it is a major challenge. She said that the role of the National Environmental Policy Act (NEPA) is critical, and that the task force has issued a draft guidance document on incorporating climate change into NEPA, which is open to public comment. She said that the government needs to adapt NEPA to adaptation, noting that climate change challenges NEPA to incorporate flexible, forward-looking approaches.

One participant asked whether anyone is thinking about how the country will adapt to mitigation. Blair replied that the task force is not looking at adapting to mitigation, although it does emphasize the need to consider the links between mitigation and adaptation.

Maria Carmen Lemos asked whether there is an inventory of adaptation actions and, if so, whether there is a focus on understanding what would be the no-regrets actions. Blair said that the best inventory of adaptation actions was published by the U.S. Government Accountability Office (GAO) in September 2009 using data up to May 2009. (The report can be found only on the GAO website.) The task force has been considering whether to

[2]*Adaptation* is defined by the Intergovernmental Panel on Climate Change as "adjustment in natural or human systems in response to actual or expected climatic stimuli or their effects, which moderates harm or exploits beneficial opportunities." *Resilience* is defined as "the capacity of a system to absorb disturbance and still retain its basic function and structure.'" See http://www.whitehouse.gov/sites/default/files/microsites/ceq/20100315-interagency-adaptation-progress-report.pdf [accessed September 2010].

update that report, but there are questions about how to define its coverage. For example, should everything the Federal Emergency Management Agency (FEMA) does be considered part of climate change adaptation? She noted that there are political choices about what is counted, predicting that the task force is not likely to go in the direction of publishing an inventory.

Kristie Ebi asked about the process for setting the priorities and who finally gets to decide. Blair replied that she cannot identify a formal process for priority setting but that there are a few key priority issues: water, coasts, health, and urban systems. She noted that there are now 12 working groups and that prioritizing across these areas is a really hard challenge.

Roger Kasperson pointed to the continued lack of serious social science input and asked how many social scientists are on the task force's five working groups. Blair said there were few scientists of any kind, except in the science working group that Claudia Nierenberg leads; the task force is dominated by practitioners. It is not defining a research agenda, but rather is trying to use what the sciences can offer today. Blair said the task force is probably not adequately addressing the social science issues, but neither is it addressing the physical science. She said they do not have sufficient guidance from social science on how to engage people, adding that there is only limited social science knowledge available to them. Other participants expressed differing judgments about how much the social sciences could offer to adaptation planning.

Nierenberg said that in the science working group, there are a lot of people from the human dimensions community and more from the program management community. This group is gathering available knowledge about natural disasters and communication. The working group's task, however, concerns moving the science enterprise closer to decision makers' needs, so it is focusing on coordination mechanisms, such as deliberative processes. Jacobs pointed out that the task force is designing an adaptation strategy but not yet a program. The Global Change Research Program (GCRP) is supposed to have been connecting science to adaptation for 20 years, but this has not been a priority before. She said that this is now a very significant part of the vision being developed in OSTP, and that it will be a point of discussion with GCRP.

Thomas Dietz asked if there is a way to have the agencies study the impacts of the actions they are taking. He also asked whether the new concept of the national assessment involves only a process, or if there will also be publications from it. Blair emphasized that agencies should evaluate impacts of their programs and that the task force is relying heavily on learning from past work. It has looked at city plans, the actions of foreign governments, and other sources to develop an approach for the federal government. She said that the task force may even have been too reliant on that sort of work.

Kasperson asked if the national assessment was going to do a comprehensive risk assessment with metrics of lives saved. Jacobs replied that it will take a risk-based approach. Ebi commented that stakeholders have different sets of priorities, a situation that raises issues of communication and capacity building. She said that if science alone is used to make decisions, there will be repercussions. Kasperson noted that a good risk assessment would not be just science. Cohen said that terms like "adaptive management" and "risk-based decision" have different meanings to different people, so that trying to apply approaches defined in such a way is problematic. He said that a conversation is needed that exposes all the mental models and questioned whether the national assessment process could put such terms out to the public so they can be discussed and defined. Blair said that the adaptation task force will not recommend national priorities or a comprehensive risk assessment approach to assess priorities. She said it was not going to preclude that conversation, but she added that the question of who organizes the conversation is an interesting one. Considering that so much about adaptation varies by geography and other factors, she questioned whether the right place to have such a conversation is the federal government.

Jacobs noted that one of the first planned workshops will be on vulnerability and risk assessment criteria for use in the national assessment. She expected that none of the major issues will be resolved in the short term but pointed out that the national assessment is a long-term process. She said the assessment will not stop writing reports. It is required to do that, and the next one is due in three years. The point in her presentation was that it is not doing the assessment just to write a report; it is intended to inform decisions. So if the assessment is asked how energy, water, and coasts intersect, it wants to be able to answer. The adaptation task force's science working group is looking at capacity mapping, to find out where agencies have capacity, what the key components of an information system are, and determining who has the needed capacity.

Moser stated that the first national assessment was not "owned" across all the agencies and asked whether the new paradigm would be broadly owned. Jacobs replied that the leaders of the assessment are trying to create an environment in which all the agencies see it as in their interest to be part of the process. So far, there has been no pushback. Several agencies will contribute, even though there is no budget item for this. Still, it will take time to make the assessment happen. Blair added that GCRP has only a limited set of agencies and that the adaptation task force has a much broader set, which includes all the key agencies in the process. The adaptation group wants to harness the value of GCRP but to engage a broader group of agencies and people.

Ian Burton commented that it is exciting to hear the recognition that the country is at the beginning of something that is being approached with humility. He said that one can set priorities within sectors and localities, even without getting everyone to sing from the same page. He concluded with the comment that adaptation must itself be an adaptive process.

7

Place-Based Adaptation Cases

LESSONS FROM THE RISA EXPERIENCE

Caitlin Simpson and Claudia Nierenberg[1]
National Oceanic and Atmospheric Administration

Claudia Nierenberg briefly described the Regional Integrated Sciences and Assessments (RISA) Program. Starting in the early 1990s as a "human dimensions" program in the National Oceanic and Atmospheric Administration (NOAA) in a climate research program, it began by looking in particular places at the relationship between knowledge about climate, particularly the El Niño–Southern Oscillation (ENSO) phenomenon, and the needs of decision making. The program was intended to assess climate-sensitive issues regionally and to teach NOAA how to build knowledge systems for information delivery. Key questions included what the critical issues are and how they are identified, what is known and needs to be known, how knowledge needs change over time, whether enough is known for effective decision making, and how to maximize social and economic benefits.

Each RISA regional team offered lessons in coordination. The coordination issues that immediately arose involved linking federal agencies with each other and with state and local governments in the region. The program offered insights into how to coordinate around outcomes, and in

[1]The presentation is available at http://www7.nationalacademies.org/hdgc/Lessons%20from%20RISA%20Experience.pdf [accessed September 2010].

more recent years it has addressed methods for evaluating the success of the program from the participants' points of view. An unexpected lesson in coordination was the value of coordinating between university and agency experts. This coordination increased the credibility of the information coming out of the program. It also proved to be very important that the researchers were stakeholders, that is, that they were located in the region and had a commitment to it over a long period of time, enabling them to work with stakeholders through implementation phases. Nierenberg said that the program was always intended to inform the development of something like a climate service. The program leaders originally imagined that the RISA teams together would, with partners at all scales of government, develop an overall research agenda that would advance knowledge for climate adaptation. The RISA program has made important contributions in this area and should be looked to as people work to broaden coordination on an adaptation research agenda.

Caitlin Simpson said that the RISA research teams are seen as providing for bottom-up, flexible responses to regional issues rather than for decisions governed by NOAA from Washington. Experience has shown that residence of the team in the region is important for monitoring change over time in physical conditions, land use, and stakeholders' perceptions, as well as for improving stakeholders' ability to interact with climate scientists. Attention to social context and to the evolution of technology have proved increasingly important as well. Each RISA team got to identify critical issues for its own region. It was important for the centers to have expertise in a range of climate time scales, from paleoclimate through models of seasonal to interannual climate variability and change. This range of expertise resonates with resource managers, whose interests are also in various time scales.

Another lesson was the importance of integrating physical and social sciences. The social sciences have been underrepresented in many RISA centers, and this problem continues, but recent calls for RISA proposals have increasingly emphasized the need to integrate the social sciences—not only to assess climate information needs (e.g., in relation to downscaled climate models), but also for analysis of vulnerability, evaluation of impacts, evaluation of RISA tools and processes, and consideration of decision-making contexts.

Simpson said the program now stresses an evaluation component from the start of projects. This includes an assessment of who the stakeholders are, what their knowledge levels are, and so forth, as well as reassessment over time. The program has learned that it is critical to have a core integration structure for network building, research coordination, and ensuring stakeholder influence on the priorities for the science agenda. Stakeholders like a central place to go where they can look at a range of climate information. Some of them also see RISA teams as a way to link to federal agencies to look at a range of issues, including agriculture, wildlife, and water.

Simpson said that from the program's viewpoint, the use of climate information for Western water issues is moving forward most quickly. Water managers have become more interested in a range of climate information, including projections of stream flow, and are interested in using paleoclimate information to frame their interpretation of the information. Some water managers want sophisticated training in regional modeling and downscaling. The program has often seen droughts as opportunities for talking about climate change and the implications for planning.

The RISA Program has done pilot work with climate extension specialists in Arizona and Florida, who work with agricultural extension specialists. The South Carolina RISA is now working with a coastal climate specialist and sea grant extension on coastal issues. Regional networks are emerging, involving university-based research teams, a set of regional climate centers, state climatologists, and federal entities, including U.S. Department of the Interior centers specifically.

An emerging area in the program involves experimenting with visualization tools, scenario planning, and other methods to communicate information that includes uncertainty to stakeholders. For example, in the Colorado River basin, there are various stream flow projections for mid-century. Several RISA teams are working together to compare models and their projections for the basin and to communicate the information in the face of their differences. Another emerging area is water and energy. Simpson ended by emphasizing that social science is critical to the RISA teams, noting that the program is working harder to identify vulnerabilities and evaluate outcomes.

Nierenberg added that one of the biggest lessons of the program so far is how much time it takes to establish relationships and get them to evolve.

In the discussion that followed the presentation, Helen Ingram said that even if few social science experts were initially involved, the program was built on social science ideas and questions. RISAs embody a social science notion of relational knowledge, which comes from communities of practice. They have thus become an important social science experiment with changing incentives for the participants, especially the academic ones. The program gave them a reason to care about what information users want, by providing funding for staff support through university teams to write newsletters and do other outreach. It built bridges between researchers and federal and local agencies and gradually attracted more social science researchers to work on these kinds of issues. She emphasized that it takes enormous patience to establish relationships of trust and collaboration among people who lack experience in talking to each other.

Maria Carmen Lemos asked how the RISA experience is informing the NOAA Climate Service (NCS) and how the service would relate to the RISAs in the future. Nierenberg said RISAs have had a profound influence on the NCS, as has the National Weather Service. She said there is now debate within the federal government about how the NCS should relate to other agencies and how it should be influenced by citizen contacts. Simpson added that the regional services component of the NCS will draw on the experiences and main lessons of the RISAs.

Thomas Dietz asked if the RISAs should also do research on the processes of working with stakeholders. Simpson said that this is very important and is actually already occurring. The program wants to look for innovative ways of working with stakeholders. The most innovative proposals in that regard in the last round of competition fared better. Nierenberg said that a difficult issue is how to cooperate across agencies as they work to diffuse that knowledge.

URBAN CLIMATE ADAPTATION PLANNING: LESSONS FROM THE GLOBAL SOUTH

JoAnn Carmin[2]
Massachusetts Institute of Technology

JoAnn Carmin began by saying that her comments may be reiterating what has already been said, but on a different scale. Her research concerns urban environmental governance: how cities make sense of climate impacts and how that links to action. The conventional wisdom about urban adaptation is that

1. the science has to be perfect, or cities can do nothing;
2. cities will not act without external incentives (carrots and sticks);
3. cities will do nothing without additional capacity;
4. public participation is essential always and often; and
5. all innovation comes from the global North.

Her research, which so far involves case studies in low and high middle-income countries, tests these items of conventional wisdom.

She has found that although scientific projections are important, disasters have been catalysts for planning in the global South. If climate projections are borne out by events, local champions often run with the

[2]The presentation is available at http://www7.nationalacademies.org/hdgc/Urban%20Climate%20Adaptation%20Planning-Lessons%20from%20the%20Global%20South.pdf [accessed September 2010].

experience. Some cities also see that advancing an adaptation agenda offers them an opportunity to be regional, national, or global leaders. Also, they do not see adaptation as something additional: climate adaptation is seen as fitting with their other priorities and as a means to advance their existing goals.

To sustain adaptation initiatives, buy-in is needed across the city and across old battle lines. People in the cities go to conferences, conduct local research studies, and look to universities and research institutes to extend local capacity and generate local knowledge. They build networks of government and research personnel and develop research agendas as a means of exchanging ideas and knowledge. They also seek out opportunities to link with cities that are similar to themselves (e.g., other coastal cities) so that they can share insights and compare experiences. Finally, cities find that they can generate greater commitment by linking adaptation to ongoing programs and by integrating climate considerations into routine activities.

Carmin stated that, with regard to science, many cities are doing model assessments with existing tools and data at a relatively low cost. Even a little local knowledge is very useful. Some cities that rely on outside consultants have been getting boilerplate reports that are insensitive to local context. Assessments are regarded as a critical step in the adaptation process. However, they do not always set the priorities for action. In some instances, they are used to legitimate action that city leaders already want to take and to demonstrate to constituents that this action is appropriate.

Public participation is not always well integrated. In the cities studied, nongovernmental organizations are not initiating these processes, and few are involved. Although a participatory process is seen as important, it is secondary to city engagement. However, some cities are testing such approaches as community-based adaptation and are drawing lessons on how to engage residents more broadly from these initiatives.

Carmin then discussed the disconnect between the five items of conventional wisdom listed above and the lessons from practical experience (see Table 7-1). (1) On the need for perfect and comprehensive science, Carmin found that cities need baseline data and want downscaled projections but often are able to work with what they have. (2) On the need for external incentives, she is finding that leading cities are taking action endogenously. (3) She is finding that although there is a need for additional capacity to initiate and sustain adaptation efforts, especially a need for funding for large infrastructure projects, cities also tend to be very resourceful. (4) Although public participation is important, there may be multiple ways to think about and develop it. (5) The assumption that the wisdom lies in the North is incorrect. Leaders in cities in the global South seek out relevant

TABLE 7-1 Conventional Wisdom Versus Practical Experience with Climate Adaptation Among Cities in the Global South

Conventional Wisdom	Practical Experience
Science needs to be right and assessments comprehensive.	Leading cities initiate action with baseline data and continue to review and expand local projections.
Cities will not take action without external incentives.	Leading cities are internally driven and find ways to link adaptation to ongoing goals and activities.
Initiating and sustaining adaptation requires additional capacity.	Cities need additional support, but also are resourceful.
Public participation is essential.	Participation is important, but seen as an element to be sequenced.
Wisdom lies in the North.	Leaders seek out and adopt relevant ideas, but also innovate and generate knowledge.

SOURCE: JoAnn Carmin. Used with permission.

ideas from the North, but they also are developing knowledge and being innovative.

Carmin's research findings have several implications for U.S. policy: (1) Cities want downscaled modeling and it would be helpful if they have these tools, but they can still proceed without it. (2) Money is being marshaled to do climate risk assessments, but dialogue is needed on what to include in them, and they need to be focused on information critical to decision making. (3) There is a need to emphasize links of adaptation to existing goals and priorities, such as sustainability and development. (4) It is important to ensure access to relevant information. There is also a need to promote measures that extend local capacity, such as university-municipal partnerships. (5) Public participation should not be promoted simply for its own sake but should emphasize critical points of engagement. Policy should be open to alternative approaches to public participation. (6) Cities are sites of initiation and of implementation and must be engaged in the early stages of planning processes. (7) Cities also are sites of innovation in adaptation. There is a need to foster multidirectional exchanges between the North and the South.

In the discussion, Stewart Cohen reported on the Columbia Basin Trust, a model from British Columbia that works to enable small towns to put climate change adaptation into their community plans. A competition offers funds for doing this, for example, by paying for planners to enable local governments to include adaptation in community plans and experts to organize in-service training for local government officials. In this model, scientists enable the local governments to act for themselves. Even these small towns can initiate action if there is a champion at a higher organizational level who can bring funds to the process. The trust spends about $30,000 per year per community.

Richard Andrews was interested to hear about the main barriers to initiation in the leading cities compared with other cities. Carmin said that she cannot find a city that is doing nothing. She added that the level of buy-in matters. If there is a powerful mayor who steps in, progress is faster than if the initiative comes from city departments. She noted that old battle lines can also be barriers to action: urban planning is often an outlier agency, and competition among groups can also hinder initiatives.

Neil Adger asked whether the city planners involved in adaptation are also involved in issues of reducing emissions and changing urban form. Carmin replied that urban planning has little influence in many of these cities. Having broad-brush city plans and strategies is a barrier to real progress because such plans are not actionable and give a false sense of accomplishment. She noted that, in most cities, mitigation and adaptation are understood as two separate things. For example, in Quito, Ecuador, mitigation is related to air quality and policy about pollution, which is an entirely separate issue from flood management and other adaptation issues. Clean Development Mechanism (CDM) projects are given priority in many cities because they are funded.

Peter Banks asked whether population growth is a more pressing issue than adaptation for the cities. Carmin responded that the issue is not growth versus adaptation; rather, there is a need to think about adaptation in planning for growth, especially with new development. She was not sure whether climate science is able to deliver what cities want to know.

Hassan Virji commented on initiation in Asian cities. In Bangkok, he said a local nongovernmental organization is leading the action and thinking about the entire future of the city, in which climate change is one of many stressors. In Shanghai, activity is more top-down, with strong involvement of developers. In both cases, downscaled climate models are irrelevant. Moreover, such models will not be available in the next 20 years. In Hanoi, downscaled model results were used to get a large loan to pay for barriers against flood surges. Still, more rain is expected, and the floods from the rain will hit the poor.

Susanne Moser asked Michele Betsill whether what she had learned about mitigation was different from what Carmin learned about adaptation. Betsill said it seems that adaptation has been harder to initiate in the

North than in the South, probably because of differences in vulnerability issues. In the United States, she has not seen cities asking for climate models, but rather for answers to questions about the economic and social effects of policies.

Roberto Sanchez-Rodriguez pointed out that any new built environment will operate under changed climatic and social conditions. He said that in addition to climate science, social science is needed to improve planning, especially in the South. Many cities have an infrastructure for planning, but it needs a lot of assistance.

Christopher Farley commented that, in the North, leadership is needed from both mayors and city managers. He observed that the cities that get involved in adaptation already tend to have mitigation plans and see adaptation as an important addition. They see the value of doing adaptation and mitigation together when they are complementary (for example, with water conservation). However, cities have different foci (e.g., water in Phoenix, sea level in Miami). He concluded that it is not possible to offer a blanket statement about whether mitigation and adaptation are treated differently.

CLIMATE ADAPTATION: FROM STORIES TO TOOLS . . . TO ACTION

Amy Luers
Google

Amy Luers reflected on some of the key workshop questions on the basis of her own experience in California with a climate policy that has an adaptation focus and on her work at Google on the role of information and technology in the context of knowledge systems. She noted that there is increasing discussion about adaptation among policy makers, but very little among the general public. She provided some insight on public communication in this area, using summaries of trends in google searches. She reported that Google searches for climate + change + adaptation are increasing in number and are more common than climate + risk, climate + vulnerability, or climate + resilience. However, compared with searches for "global warming," the number is minuscule.

After California passed Assembly Bill 32, when the state put a mitigation plan in place, people started asking what they should do about adaptation. She said that it is difficult to articulate the climate adaptation challenge to nonscientists. The mitigation challenge is relatively easy to explain, as the need to reduce greenhouse gas emissions can be explained with such concepts as "stabilization wedges." With adaptation, people want to know what specific problem they need to solve, what the options are, and what is at stake so they can debate the options. The bigger challenge, however, is to build the knowledge, institutions, and culture to support adaptive management. The problem can be articulated in terms of information scarcity: there are lots of information sources, but there is the need to acquire information, disseminate it, and make it actionable. Luers said that a more distributed and participatory approach is needed for climate change science. Information and communication technologies (ICT) have created the web structure, but it has not been exploited. There are many data portals with Earth observations and social networks to enable social learning. However, the most pressing constraints are not in data or technology, but rather are institutional, cultural, and economic. The disaster community is well ahead of the curve on how to organize ICT in an innovative way to provide more rapid information for responses to disasters. One example is of a community "crowd sourced" map that was created within a week after the cyclone in Myanmar.

Luers raised the following key questions: Today in the information age, how do people access and gain trust in climate information? Is there a role for the web and wiki environments to connect with scientific assessments? How can institutions be developed to support ICT on climate adaptation? How can the pieces for an adaptive and responsive ICT system be put into place in an unplanned world? Can adaptation systems follow such a model?

In the discussion, Cohen commented that the disaster community uses information technology to display observations, which can be confirmed. By contrast, scenarios and futures cannot be confirmed in the same way, even though they are easy to visualize and even though they really get people's attention in communities. They can visualize a house in a flood plain, causing its value to decline, but the visuals are art, not science. Luers said that the field is already heading in the direction of increased use of visualizations and the like, and the question is how to design these outputs. The community has not gone far in figuring out how to use the available tools. She said she has been advised both to make downscaled projections available and not to do so. She noted that it is possible to block data representation at really small scales to avoid misinterpretations.

Dietz asked if there is research on whether these tools are disseminating better information, or just helping people confirm their preconceptions. He also asked whether web-based tools allow data acquisition through surveys and the like. Luers said that google.org has a program to look at the use of cell phones, etc., and universities are studying the use of open-source platforms. She agreed that there are many social science questions not being asked about how people use information tools and about the relative strength of influence of information from different sources.

8

Adaptation and Natural Resource Management

ADAPTING TO CLIMATE: LEARNING FROM THE CAROLINAS WATER RESOURCES SECTOR

Kirstin Dow[1]
University of South Carolina

Kirstin Dow began by noting that the Carolinas Integrated Sciences and Assessment (CISA) Program is one of the Regional Integrated Sciences and Assessments (RISA) Programs of the National Oceanic and Atmospheric Administration (NOAA). Dow started by stating CISA's mission, which is to improve the range, quality, relevance, and accessibility of climate information for decision making and management of water resources in North and South Carolina. Communities in those states as are willing to contribute to the costs of statewide information systems and are looking for information at a geographic scale matched to their decisions. Many federal and state agencies are undertaking studies, but there is a lack of coordination and communication. There is also an issue of the capacity of organizations to keep pace with interest, new research efforts, and data requests. CISA is also receiving many questions about mitigation policy. In the Carolinas, some federal stimulus money went to hiring city sustainability coordinators, who are looking mainly at mitigation but are also addressing adaptation by improving energy efficiency in low-income housing.

[1]The presentation is available at http://www7.nationalacademies.org/hdgc/Adapting%20to%20Climate-Learning%20from%20the%20Carolinas%20Water%20Resources%20Sector.pdf [accessed September 2010].

Drought is a chronic issue in the Carolinas. There have been worse long-term droughts in the past than in recent years, but the population has doubled, so droughts are more serious.

The RISA project in the Carolinas started at a time when the Federal Energy Regulatory Commission (FERC) was requiring relicensing of dams and also right after a major drought. FERC wanted a low-inflow protocol based on local information and acceptable to the localities. Geographic information system data were available on various drought parameters at a fine scale, although the data are collected with different frequencies. The project tried to determine what kinds of local information managers needed. County-level information was very popular, even though it was not meaningful in an environmental sense. On the trade-off between accuracy and precision, Dow noted that few water managers understood that accuracy is greatest where the gauges are. However, those are not the places where they most wanted accurate information. Maps of projected sea level rise rarely come with data showing what climate scenario they are based on or show uncertainties in the digital elevation models, even though these details are associated with huge differences in the number of people who are at risk. The CISA project has worked on improving visual communication and representation of uncertainty in climate maps, drawing on advances in "cognitive cartography."

The project is working to estimate vulnerability to drought. The cost of drought has been estimated at $6-$8 billion per year, although the data for generating those numbers are quite thin. These estimates cover losses only in certain sectors, differ across states, and do not always correlate well with the losses covered in the local newspapers. Also, little is known about low-income populations. Some low-income jobs are lost first in a drought (for example, landscaping, car washes, pool maintenance), and no one knows how many people are involved. Also, the prevalence of shallow water wells among the rural poor causes assessment problems for drinking water effects. Analyses have not paid much attention to the effects of drought on household budgets, such as through water bills. Dow reported that research on visual representations of uncertainty in drought maps was scheduled to begin in summer 2010.

In the discussion, Dietz suggested that much more needs to be learned about how to present information about risk, uncertainty, and distributional impacts. A participant also noted that using the web to collect impact information is problematic because people do not always enter the information, and managers lack the capacity to use it well.

KNOWLEDGE, NETWORKS, AND WATER RESOURCES

Helen Ingram
University of California, Irvine

Helen Ingram began by citing the 2008 National Research Council (NRC) report on the NOAA Sectoral Applications Research Program, which expressed concern that social science research resources at NOAA would be drained. She sees that concern as increasingly serious now. She mentioned that the NRC report also criticized a decision support "loading dock" model that allowed only one-way rather than two-way communication, noting the need for decision support to be more iterative. Her comments also draw on the U.S. Global Change Research Program's Synthesis and Assessment Product 5.3 (U.S. Climate Change Science Program, 2008), which contained significant social science content, including several good case studies, and also emphasizes the need for an iterative process of climate information translation.

Ingram emphasized that knowledge production is relational, so it is important to ask who is involved in it and what perspectives are included. She also said that the social sciences often fail to give attention to the effects of physical phenomena. Water is place-based and is understood as part of local identity; these aspects of water are not handled in markets. She said that the idea of best practices, because it is generic, often means imposing an inappropriate model on a particular situation. Knowledge for adaptation has to flow across disciplinary boundaries and has to be developed together by users and scientists; that is the only way people will come to "own" knowledge. In the relational concept of knowledge, verification is less important than salience, trust, and legitimacy. Water managers want to know what people think, but the tools and terminology they use tend to exclude people. Building legitimacy takes a lot of time and effort. Trusting relationships, once broken, take a long time to repair. With water management, there is a long history of mistrust. Knowledge networks require a special kind of leadership that involves individuals and organizations who serve as boundary spanners, conveners, and brokers, understanding and respecting different perspectives.

Ingram said that coordination is a permanent problem, not easily solved. Agencies are better at talking about things and creating institutions that disperse the risk of tough solutions, appearing to resolve problems while in reality postponing judgment rather than actually making new, coordinated decisions. The 1965 Water Resources Planning Act (42 U.S.C. §§ 1962-1962d-3), for example, was supposed to end coordination problems over water in the federal government. It set up a Water Resources Council and River Basin Commissions aimed at getting agencies at the federal and regional levels to speak with one voice. These coordinating bodies never functioned as planned and are now mainly disbanded. There is a long history in water of visionary legislation that makes big promises that have led to weak performance. Coordinating bodies in general have not had a good track record of accomplishment, although they deflect political heat. Ingram said that agencies are good at appearing to put problems to rest. Another example is the Synthesis and Assessment Product process. Ingram and her colleagues worked for 2.5 years on SAP 5.3 and, when it was released, there was little sign of immediate policy maker or media interest.

The water community has failed to mobilize for change. This is due not only to inertia, but also to opposition from entrenched groups opposed to losing their responsibility and control. The professionals in the field want to keep water as a low-visibility issue. Ingram mentioned issues of equity in water supply and raised the issue of whether production of data on inequity would create a constituency for that kind of information.

She considered ways to transcend inertia. One is to bring in new constituencies and generate new alternatives, for example by building new networks, such as the RISAs. Another is to pay more attention to framing stories and narratives. The right frameworks could overcome differences and heighten stakeholders' sense of shared destiny.

In the discussion, Bonnie McCay said that much of what Ingram concluded also applies to marine fisheries. The idea of boundaries is very apparent there and is more usable to social scientists than the idea of coordination. Ingram agreed that boundary organizations and experiences are very valuable, more so than coordinating committees.

Caitlin Simpson asked about diffusion of innovation among water managers. Ingram said that many municipal agencies want to be on the cutting edge of technology and therefore want people who can speak the latest jargon—regardless of whether it is used in decision making. Keeping up with the Joneses is an important motive in the water sector. She wondered whether enthusiasm for planning can stand in for having new people. People get swept up in ideas even if they are old ideas in new guises. Perhaps innovation is prompted by new framing, even if the framing is not very original.

Ashwini Chhatre underlined the issue of conflict: every intervention has multiple outcomes that are impossible to anticipate and can create conflict. Coordination and collaboration assume that conflict can be resolved or smoothed over by "reaching consensus" (which can merely sweep conflict under the rug), rather than using conflict to make explicit the differences in values and impacts that are inherent in the choices. There is a need to build more inclusive knowledge networks and set up processes in which trade-offs can be made. Some of the best changes have come through conflict. However, governmental systems for dealing with these things do not manage conflict well. Water is a good example, in that it does not match political subdivisions. Ingram said that the way to encourage innovation may not be to reduce conflict but to increase the numbers of participants and help them understand where their interests lie. She does not think it makes sense to manage conflicts with coordinating committees. People from different disciplines think they understand and own the water issue, but in fact each one sees it only from a single perspective. Water is a paradigm case for building knowledge that is more inclusive. Ingram also noted that conflict has led to a lot of positive progress in water management, for example, in the 1970s.

Susanne Moser said there are very consistent stories across domains and countries: the same lessons repeat and management systems never learn. She said it is a waste of time to go down the same road again. More important is to figure out how to get people to learn. Ingram said that new ideas quickly become rationalized into management systems, with nothing changing—unless constituencies can be mobilized that will have an enduring interest in overseeing the implementation of new ideas.

Roberto Sanchez-Rodriguez noted that people do not connect sectors and address issues of broader social well-being, which would bring in more actors. Ingram said that water resources research may have become too bounded.

ADAPTATION TO CLIMATE CHANGE IN THE ATLANTIC SURFCLAM FISHERY: AN EXEMPLARY OR CAUTIONARY CASE?

Bonnie McCay[2]
Rutgers University

Bonnie McCay began by speaking broadly about responses to climate change. She asked whether human responses are negative feedback that control the systems or positive feedback that amplify preexisting trends. She said that positive feedbacks in fisheries might be counteracted by changes in property rights regimes. Although marine fisheries are not on the list of critical issues for climate change adaptation, climate change does in fact significantly affect fisheries. Fish and shellfish are experiencing changes in fertility, growth, and mortality and moving in apparent response to warming trends in the oceans. Complicating the study of how fisheries adapt to climate change is the fact that the fisheries sector is already organized in response to other mandates, and climate change is a new force in the system.

---

[2]Presentation is available at http://www7.nationalacademies.org/hdgc/Case%20of%20Adaption%20to%20Climate%20Change.pdf [accessed September 2010].

The responsible federal agency, NOAA's National Marine Fisheries Service (NMFS), monitors the system, but climate change is altering it. McCay briefly described a research program of the National Science Foundation that she is involved in, called Coupled Natural and Human Systems (CNHS). Her project, in which she is working with oceanographers, economists, and ecologists, examines the surfclam industry. Surfclams are caught from the mouth of the Chesapeake up to Georges Bank, off the coast of New England, but the waters off New Jersey are currently the main fishing grounds. Big boats dredge the clams, which are then taken to factories and made into clam chowder and other products. The surfclam industry was the first to be privatized with a tradable quota system. Instituting the system reduced the tonnage of boats by half almost immediately, with rapid consolidation of ownership, suggesting that there should be low transaction costs associated with collective action to adapt to climate change. Some evidence of that is the fact that the industry was a leader in developing collaborative and industry-funded research with NOAA, but it is as yet uncertain whether or how it will adapt to the effects of climate change.

Surveys began to show some trouble in the fishery around 1998. Clams began dying off at the southern end of the range, off the Delmarva Peninsula, coincident with warming in the ocean. Since 2002, there has been a dramatic decrease in catch per unit of effort not only there but also to the north, in New Jersey waters; this apparent decline in the abundance of clams has been interpreted as possibly due to climate change rather than to fishing pressure. McCay's research is seeking to find out what is going on in the marine system affecting the clam population, not only through coupled hydrodynamic, biological, and genetic modeling but also by documenting the responses of industry, scientists, and managers. The clams do not seem to be moving north, hypothetically because their larvae are not making it across the Hudson Canyon, one of the deep submarine canyons of the region. Some fisheries have closed; some have declining productivity. The research has examined a range of responses—moving to more abundant clam areas, fishing more intensively, switching to other species, shifting processing factories northward, and taking business buy-outs.

The research is to see whether the socioeconomic and policy responses amplify or compensate for the environmental changes. What can be done? The National Marine Fisheries Service and the regional fishery management council—or the industry voluntarily—could close certain areas temporarily with a rotating fishery, protect the remaining viable clam populations with lower quotas, or change size limits on the clams. These things have not yet been done. The issue is not officially discussed in terms of adaptation to climate change. Scientific uncertainty about the role of climate change adds to tremendous resistance to change in practices, given the existence of well-developed institutions and a system of privatized fishing rights that demands as much asset security as possible. The fisheries businesses have become deeply invested in their tradable quotas, mortgages, etc., and lobby passionately for no change.

The tradable quota system has left fewer foxes guarding the henhouse but leaves open the question of whether they will be good stewards given the premium on predictable and consistent yields (in this case furthered by the close tie of the industry to large retail and institutional markets). The current management system has well-defined rules, such as that one must manage the stock throughout its range and therefore is constrained from applying different management rules for different places. It also requires making decisions based on the "best available" (i.e., peer-reviewed) science. McCay's research group is working with NMFS scientists and industry to bring climate change into the stock assessment process, in order to get the best available science into the decisions. McCay is interested in how that happens and also how the industry understands the situation. The industry has not yet overcome its collective action problems, even though there are few actors.

In the discussion, Michele Betsill suggested that the well-developed institutions are a barrier because they support certain interests and keep value conflicts from being confronted. She suggested bringing in climate forces that conflict. McCay noted that privatization has led industry to make large investments, with the result that it does not want any changes in the quotas or the other rules for doing business. The inflexibility of the system is enhanced by the environmental community, which has restructured the fisheries management system to make it very precautionary under uncertainty. Climate change will increase the uncertainty, and by law that will require the management councils to reduce the catch. So fishers have an incentive to keep climate change out of the system. The management system needs to change, but that is not likely.

One participant commented that social science tends to be deployed to identify narrow solutions but not to look at the larger social context. McCay responded that social scientists ask for funds to study the social context but are unsuccessful. Then if the science is not there, it is not used. Agencies give money through the political process, but the social sciences are rarely called upon.

Neil Leary pointed out that the behavior of this fishery runs counter to what economic analysis would predict. The fishery has been privatized, yet the incumbents are resistant to change, even though they understand the problem. He asked why there is reluctance to use available information about climate change. McCay said that there are other economic factors at play and that her group is studying the question. For example, one of the largest firms has a business plan based on the reopening of clam beds to the north, which carry the risk of a toxin and have been closed pending the development of a method to assess the risk.

Adam Henry noted that values can change and lead to social transformation and also that social networks can play a role in value and belief change. He asked if anyone has examined a collaborative process in which value change has occurred. McCay said that efforts have been made to change values toward privatized rights by bringing people from the North Pacific to events to try to change values. But people are resistant. She said that the greatest changes occur when new people move into new positions. Continuing the discussion of values, Moser noted that well-facilitated dialogic processes can open people to listening to those with different values. Kasperson pointed to historical quick change in values about nuclear technology and environmental protection.

ACCESS, ARTICULATION, AND ADAPTATION TO CLIMATE CHANGE

Ashwini Chhatre[3]
University of Illinois at Urbana-Champaign

Ashwini Chhatre noted that social scientists can bring a unique perspective to climate adaptation. Historically, he said, people have not treated the climate as stationary. They have responded to climate as it came and tried to draw lessons from history. There are two kinds of data about climate change processes: coarse resolution climate models (he said that they will never be fine-scaled enough to help anyone in deciding about a local problem), and fine-scaled social science. Climate models get worse as the scale gets finer, but social science data become better as the scale gets finer. There is no a good coarse-scaled social theory, and fine-scaled expertise exceeds global expertise. This situation allows social science to make headway on adaptation, which is largely local. Social scientists can draw on knowledge of related phenomena and insights from various theoretical perspectives to arrive at a social science synthesis.

In Chhatre's research perspective, adaptation is first the property of a system. Adaptation practices are localized responses to risk, including climate variability. Institutions are mediators that structure responses, making some responses possible and others difficult. Landscapes and institutions are mutually constituted and follow coevolutionary trajectories. His research project has looked at how institutions enable or disable adaptation

[3]The presentation is available at http://www7.nationalacademies.org/hdgc/Access%20-%20Articulation%20and%20Adaption%20to%20Climate%20Change.pdf [accessed September 2010].

practices locally, examining many cases. The practices include diversification, mobility, storage, pooling, and exchange. Institutions exist at multiple scales and enable or facilitate certain kinds of practices: national laws and community civic institutions matter most.

Complex adaptive systems are emergent properties of adaptation practices. In the Indian Himalayas, the core production zone of apples has shifted 8 km northward and 1,000 ft higher since the late 1980s. There is a clear link to climate that was not identified until the late 1990s. Daily high temperatures have increased for the first 16 weeks of the year, when it matters most to apples. In the districts where Chhatre works, income from apples dropped from 80 percent of household income to less than 5 percent, while total incomes increased. People diversified into other fruit trees that were more tolerant of the current climate and into fresh vegetables in the off-season for urban markets. This change was not uniform. Land is heterogeneous, and the shifts from apples have occurred only where the biophysical characteristics of the landscape allowed it.

This system transformation was facilitated by institutions that provided technology, know-how, and subsidies. The institutional network included university research centers, inexpensive credit through banks and cooperatives, and regulated market access. The nodes in the network have been created by the local people since the 1960s, either themselves or using institutions that were created for other purposes. The network provided access to institutions for poor and vulnerable groups (women, the poor, and low-caste groups). Information flow across institutions was driven by demand, with feedback through the institutional network operating through local elections. In hindsight, the system evolved in an adaptive way. Adaptation planning in this case involved an investment in local networks to allow for self-organization, rather than a top-down enterprise. These are investments in democracy. Chhatre proposed that this strategy for adaptation will be messy, but it has a better chance of working than top-down planning. However, for this approach to work at the system level, there needs to be access to a diversity of institutions, cross-scale articulation, monitoring and improvement in flows of information and in institutions, and infusions of money and energy into existing institutions.

In the discussion, Ingram commented that Chhatre's presentation showed how adaptation is guided by existing institutions that continue past patterns. He seems to assume that the density of cross-institution ties produced beneficial results, a result that contrasts with Ingram's experience with water management institutions in which these ties resulted only in spreading risk. Henry commented that, in this case, existing networks evolved to lead to adaptive outcomes and asked how this came about: What is the evidence? He suggested that Chhatre's assumptions appear to be based on secondary analysis of case studies and questioned whether they are reliable. JoAnn Carmin commented that institutions generally resist change, but Chhatre claims that entrenched institutions sometimes do change. The assertion that is sometimes made, that democratic voting produced these changes, is not clearly supported. It is merely asserted in preference to the counterintuitive claim that institutions resistant to change are actually the drivers of adaptability in this case. Is this really in effect a tautological argument identifying density of institutional linkages, adaptation, and democracy?

McCay asked about the broader theory underlying the case, and Kasperson asked if there is a good statement somewhere of the major theoretical approaches to adaptation and also whether Chhatre's theory is supposed to be general or specific to certain domains. Chhatre said he is trying to develop a general theoretical framework, drawing on several existing theories. He noted that his database includes cases of failures as well as success. In terms of network density, he sees the system as changing over time. He thinks research has focused too much on institutions that do not change, whereas most of the institutions he has examined in India have changed beyond recognition within 30 years. He is working toward a theory of change as opposed to correlations between static conditions. He noted that there is not a good understanding of how change happens, even though institutions have changed dramatically. The difficulty has been that the institutions that one would like to change do not change. Chhatre is working with a set of 125 cases to try to get a sense of which level of institutions are most important to change. Moser disagreed, arguing that there are multiple theories of change in different social science disciplines—policy change, behavioral change, change brought about by social movements, organizational change, spread of innovations, and others—just not one grand theory.

9

Cross-Cutting Issues in Adaptation

INTRODUCTORY COMMENTS

Roger Kasperson

Roger Kasperson opened this section of the workshop by saying that a lot of adaptation is going on, and there is a theoretical framework for that—muddling through. He said that adaptation responses vary by sector and have some specificity, so that there are no golden answers that apply to everything. He suggested that something may be said about how to avoid common pitfalls, but it is not easy for social scientists to talk about how to transform entire systems. At the end of the workshop, people should try not only to pull together what they have learned, but also to take on really hard questions, such as: How can large adaptations move forward at the system level? What advice can social scientists offer to the government about major transformations needed for the climate change problem? How can success be assessed?

Paul Stern suggested that for government, it is important to deal with practical issues and not just the major social science questions, such as what major social transformations are needed and how to get at them. Neil Adger noted that there may also be important fundamental social science questions related to adaptation, such as about processes of transformation, the evolution of preferences over time, demographic change, and relocation of settlements and economic activity.

LESSONS LEARNED FROM PUBLIC HEALTH ON THE PROCESS OF ADAPTATION

Kristie Ebi
Intergovernmental Panel on Climate Change

Kristie Ebi began by observing that the public health field has 150 years of experience in dealing with everything from slow changes in factors affecting human health to dramatic epidemics. It includes national and international organizations and institutions for identifying risks and implementing programs to reduce or eliminate the threats. Because some of the experience in this field could provide insights to improve the process of adaptation in health and other sectors, it might be useful to organize case studies on lessons learned around selected questions, for example, communicating to facilitate behavioral changes.

The public health field has extensive experience with communicating on a range of issues. For example, 40 years of experience with communication has taught that different audiences need different messages on the hazards of cigarette smoking: effective messages for young adults differ from those for older adults. It is possible to effect behavioral change, but the most effective way to do so is often not predictable, and a variety of options need to be tested to determine which are most effective. She related a story from Rita Colwell about how, in Bangladesh, a simple practice of filtering water through used sari cloth dramatically reduced the cholera burden in one region. Ebi noted that one or two individuals can make a significant difference quickly, as was the case with Mothers Against Drunk Driving. Such experiences can teach lessons about messages for adaptive action.

The public health literature identifies five prerequisites for action: awareness that a problem exists, understanding of the causes, a sense that the problem matters, the capability to influence the problem, and the political will to act. Political will is often the most significant constraint.

An adaptation option of considerable interest is the development of early warning systems based on environmental variables. A challenge with many of the early warning systems developed in public health is that they were not designed to adjust to a changing climate; they implicitly assumed a stable climate. In many cases, it will be a challenge to proactively identify through an early warning system where a disease might change its geographic range due to climate change. Furthermore, there is no definitive approach for deciding when to retire early warning systems in some places and open them in others as diseases change their range. This means that some mistakes will be made. She noted that in public health, thresholds are often constructs, not biological limits. For example, the definition of high blood pressure is based on judgments about the costs and benefits of treating people in different groups—there is no biological justification for choosing a particular blood pressure as the threshold above which risks increase significantly.

Ebi emphasized the difficulty of maintaining public health systems over time. Yellow fever was controlled in every country where control of the mosquito vector was tried, but that costs money and takes effort. Mosquitoes can reappear where there has been failure of vector control. A lesson here is that climate change adaptation can require a very long commitment.

She also noted that things can go very badly even with the best of intentions. In Bangladesh, for example, in an effort to stop childhood deaths from water contamination, there was a large effort to drill tube wells that accessed uncontaminated ground water. However, some wells were seriously contaminated with arsenic. The implication was that although there is a bias for looking for simple, single-technology fixes, it is important to understand the broader implications of implementing a technology.

In the discussion, in response to a question, Ebi observed that effective early warning systems require a process, not just a one-time design focused on identifying a threshold for action; the actions to be undertaken also need to be carefully designed and tested. Incorporating climate change into public health systems means considering what is likely to need adjustment, so that systems can be easily modified with additional climate change.

On the topic of decision thresholds, Howard Kunreuther said that thresholds are often used for regulatory purposes, even when there is uncertainty and a continuum is better justified scientifically. Ebi responded that public health has moved to an approach based on judgments of how many people could be saved by setting thresholds at different levels.

THE NETWORK STRUCTURE OF CLIMATE CHANGE ADAPTATION: VIEWING NETWORKS AS OPPORTUNITIES AND BARRIERS TO SUCCESSFUL LEARNING

Adam Henry[1]
West Virginia University

Adam Henry began by saying that climate change adaptation is an important form of policy learning, which is one of his central interests. Because no one knows which policies will work well in advance, risk management needs to be adaptive or iterative. He said that to understand policy

[1]The presentation is available at http://www7.nationalacademies.org/hdgc/The%20Network%20Structure%20of%20Climate%20Change%20Adaptation_%20Viewing%20Networks%20as%20Both.pdf [accessed September 2010].

learning, it is necessary to engage in network analysis and that a network perspective can help one to understand opportunities for, and barriers to, successful climate adaptation. In particular, networks can facilitate or hinder policy learning. They can facilitate learning by promoting coordination across sectors or scales, and they can hinder learning by fragmenting or shutting out information. To promote successful adaptation, one goal may be to create certain types of networks and avoid others. The design of adaptive institutions must account for the fact that policy networks self-organize.

Henry examined one of many possible applications of network analysis to climate change adaptation. He noted that network sampling techniques can be used to map the participants in climate change adaptation. Stakeholders can be linked to each other directly or indirectly, by physical interactions and by cognitive relationships (e.g., trust). He said that standard statistical methods of analysis are often inappropriate because of interdependencies, but other methods exist.

The definition of adaptation from the Intergovernmental Panel on Climate Change (IPCC)—"adjustment in natural or human systems in response to actual or expected climactic stimuli or their effects, which moderates harm or exploits beneficial opportunities"—implies challenges for learning. These include understanding how the system works, avoiding perverse learning, inducing learning that occurs between communities of knowledge and action, and the learning of common goals and values. Networks can affect learning by exchanging information, promoting dialogue, building trust and credibility, spreading innovation, and diffusing values, beliefs, and other cognitions. Henry also noted a common belief that networks that link diverse agents across disciplines, world views, etc., have the effect of improving problem-solving capacity. He observed that there is some support for this belief from theory and practice.

"Network" has many meanings. It is a mathematical abstraction that focuses on the relationships among agents. Networks can be studied with mathematical techniques from graph theory and social network analysis to analyze the effects of network position and structure. The nodes in networks are the agents; the group of all the agents is the boundary of the network; the links are the relationships between nodes. The set of all the links gives the network structure. In popular usage, networks are seen as a good thing; however, there are many network structures, and different structures can function differently and enable learning to different degrees. The question of which structures facilitate learning can be asked at different scales: the egocentric or single-actor scale and the macro- or network-wide scale.

The most coherent discussion of the effect of network structure on the capacity to learn comes from the collaborative policy literature (e.g.,

Schneider et al., 2003), which focuses on fragmentations of networks ("structural holes") that prevent innovation, such as when there are communities defined by dense connections within that are linked with each other only by sparse connections. This literature hypothesizes that increasing collaboration across structural holes (e.g., across sectors, jurisdictions, levels of government, ideological divisions) will increase the likelihood of successful adaptation—but there has not yet been much testing of this hypothesis.

At the egocentric level of analysis, Henry identified four hypotheses in the literature. First, the more expansive that ego (the learning agent) is in the sense of number of collaborations or links, the greater the likelihood of successful adaptation. The assumption here is that more information is better. Second, reciprocity (connections with flows in both directions and without hierarchical relationships) promotes learning. The underlying theory is that learning is a coproduction process, not one just within individuals. Third, clustering (embeddedness in triadic relationships) increases the propensity for successful learning. The theory is that redundancy improves trust and legitimacy. Fourth, if the agent goes to diverse information sources, it has a better chance of success in adaptation. This idea presumes that agents apply good ideas to their particular situations.

Henry reported on a study he conducted on 71 networks, which does not support all these hypotheses. For example, it does not support the first hypothesis and supports the second only contingently. However, network variables explain a large amount of variance in a measure of innovation. This means that something is known about the kinds of networks that adapt well.

Given that knowledge, Henry asked, what can be done to develop adaptive networks? One issue is that networks self-organize: participation in them is generally voluntary. Also, individuals may position themselves so that networks become barriers to successful adaptation. Ronald Burt's (2004) work suggests that although structural holes may be bad overall, some individual actors gain power by spanning them and so try to maintain them. The advocacy coalition framework of Sabatier and Jenkins-Smith (1993) suggests that networks form around shared belief systems, a process that limits diversity. Some of Henry's data support this model.

Overall, Henry concluded that the network perspective is useful, that network sampling and modeling methods are useful, and that more work is needed on learning within networks.

In the discussion, Henry was asked about learning in climate change adaptation networks: Does he equate successful learning with successful adaptation? Can networks can be successful at learning the wrong things? How can the lessons of experience be used to design networks for adaptation?

Henry replied that there are many evaluative criteria related to learning. For example, shared understanding is an indicator of learning even if network members do not agree on policy. Actors may arrive at a common strategy even if their values do not change. He added that because individuals span different kinds of networks, it is important to be clear about the boundaries of the networks. He said that the big empirical questions concern how to characterize networks that do well or do not do well.

THE ROLE OF URBAN AREAS IN ADAPTATION AND EFFECTIVE STAKEHOLDER-RESEARCHER ADAPTATION PROCESS

Cynthia Rosenzweig[2]
NASA Goddard Institute for Space Studies

Cynthia Rosenzweig explained that she has been working with New York City for a decade, since the first National Assessment of the Consequences of Climate Change. It was clear from the start that cities like New York have global connections as well as local roots. At the 2009 Copenhagen conference, New York participated in the mayors' summit, which was hosted by the mayor of Copenhagen. Unlike the experience at the nations' summit, the mayors had no trouble setting targets and timetables or creating adaptation plans. This was the case for several reasons. The vulnerabilities of cities are acute, with high population densities, coastal exposure in many, higher levels of heat, and poorer air quality. On the mitigation side, cities are responsible for 40-80 percent of greenhouse gas emissions, depending on how the estimate is made. Also, mayors are closer to the citizens than heads of state are. And cities have some readily available policy levers. They operate critical infrastructure (e.g., they can clean storm drains to prevent floods), and they have budgets for maintaining it. Cities also want a place at the table for funding for adaptation.

Cities also face barriers to action. For example, their leaders have other pressing issues, constrained resources, electoral concerns, and all deal with multiple jurisdictional issues. But cities are forming international linkages around climate change. C-40 is the large-city climate summit, and the International Council for Local Environmental Initiatives or Local Governments for Sustainability is playing a major role, for example, by organizing side events at Copenhagen. The intercity effort is considering questions about learning and about transfer of knowledge between cities in high- and low-income countries.

[2]The presentation is available at http://www7.nationalacademies.org/hdgc/Thoughts%20on%20the%20Role%20of%20Urban%20Areas%20in%20Adaptation%20and%20Effective%20Stakeholder-.pdf [accessed September 2010].

All this activity is leading to an intense research interest in the role of cities in climate change and in developing an effective assessment process for climate change and cities. IPCC has responded by giving cities a more prominent place in the next assessment. An ongoing assessment for cities may also be needed, parallel to the one for the countries. The United States could take the lead on this, inasmuch as it already supports 50 percent of the IPCC's budget.

The New York experience identified several challenges: (1) responding to the need for rapid, recurring assessments; (2) enhancing coordination among stakeholders, jurisdictions, and scenarios; (3) handling the uncertainty of climate information; (4) revising standards and regulations; and (5) defining and implementing the role of the federal government. Adaptation planning worked in New York for several reasons. There was high-level buy-in by the mayor and the long-term planning office above the city agencies, which was important for coordinating across agencies. An outside consultant designed the process, and a stakeholder task force attracted agencies and nongovernmental groups. Expert knowledge was separated into a technical advisory committee (Mayor Bloomberg changed the name to the New York City Panel on Climate Change), with public and private participation (e.g., involvement of the legal community). The process has not always been cordial because there are differences in interests and culture. The experts are analogous to the parents of the climate change issue; the cities are like the teenagers who are growing up and have to take it over. Rosenzweig said that for successful adaptation, it is important to set up ongoing structures like these, rather than one-time activities.

Workshop participants raised a variety of issues during the discussion: how to engage cities that are resistant to taking action, how expert knowledge can be engaged to meet the needs of the cities, how cities can function effectively within a process dominated by nation-states, what smaller cities can do, how experience can be transferred to smaller communities, how New York overcame myopia to focus on long-term goals, and whether there were critically important boundary organizations in the New York City process.

Rosenzweig noted the approach used by the Regional Integrated Sciences and Assessments Program and the fact that the National Oceanic and Atmospheric Administration has created an urban RISA. She said that urban research centers like this can be very important. She said that the New York City Panel on Climate Change provided common scenarios at stakeholders' request, with the associated uncertainties. She said every actor has to be involved and that coordination, rather than competition, must be the goal.

OBSTACLES AND OPPORTUNITIES: LESSONS FROM CASE STUDIES OF ADAPTATION TO A CHANGING CLIMATE

Neil Leary[3]
Dickinson College

Neil Leary described a project under the international SysTem for Analysis Research and Training (START) Program that produced a set of 24 case studies executed by groups of researchers and stakeholders in developing countries in extremely varied social and environmental contexts. The case studies indicated that because adaptation benefits are largely internalized (much more so than mitigation benefits), countries have a strong incentive for adaptation and it is happening in many places. Still, researchers commonly see an adaptation deficit.

The reason, Leary said, lies in obstacles to adaptation, which roughly include (1) social inertia (lack of determination or political will), (2) lack of means, and (3) the public-good aspects of adaptation. Political will or determination appears when people find a substantial threat to things they value; when reducing the risks is a priority on par with other major goals; when they can see effective and affordable options; and when they know enough about the problem to make wise choices. The case studies identified four reasons for lack of determination to adapt: one is information problems (e.g., doubts about whether recent trends are reliable indicators of climate change). A second is competing or opposing objectives (perceived risks are low, distant in time, or less than pressing current priorities. Attitudes changed somewhat when climate change was connected to climate variability and extremes and when climate change was seen as threatening things the particular people valued, such as health, livelihoods, or development. A third reason was improper or misaligned incentives that shield some actors from the consequences of risky behavior. For example, in Mongolia, after collectives were dissolved, land was treated as a commons and herders were free to graze on public lands, leading to underinvestment in improvements of the pasture and water supplies that could build resistance to stresses. Finally, lack of agency or inability to act was a significant obstacle in several case studies. Lack of financial resources was a universal problem. In some countries, poverty, degraded natural resources, inadequate infrastructure, weak local institutions, and poor governance were problems. There are significant public-good aspects to adaptation, including the needs for community development, poverty reduction, and the provision of information and cocreation of knowledge.

---

[3]The presentation is available at http://www7.nationalacademies.org/hdgc/Obstacles%20and%20Opportunities_%20%20Lessons%20from%20Case%20Studies%20of%20Adaptation%20to%20a%20Chan.pdf [accessed September 2010].

Opportunities for intervention to increase action on adaptation include strengthening the web that connects actors: strengthening nodes in the web, adding new nodes to meet strategic needs, strengthening links of knowledge to action, providing opportunities and resources to increase interactions between nodes, making connections across scales, development of programs for cocreation of knowledge, South-South knowledge sharing, and "pro-poor" development.

In the discussion, one participant underscored the importance of capacity building efforts in the South and said that START has worked to build science-policy dialogues and improve risk communication, which increases the need for social science involvement. Carmin asked whether the type of innovation (e.g., in communication or in technology) influenced the results. Adger asked about the role of poverty (including seasonal poverty) in adaptation. Ian Burton noted that cross-country comparisons are difficult for this project because it started quickly, with limited time for design. Chet Ropelewski noted that climate trends are hidden by variability and asked for expansion on how information about climate variability was useful for inducing action.

Leary said that technology was not high on the list of barriers to adaptation. However, the inherent weakness of local organizations was a major barrier—organizations that are poor, busy, or include people who do not see climate as connected to their visions are unlikely to act on adaptation. Getting support for the cocreation of knowledge was a difficult point. Poverty was an issue mainly as it related to capability.

ADAPTATION TO CLIMATE CHANGE THROUGH LONG-TERM CONTRACTS

Howard Kunreuther[4]
Wharton School, University of Pennsylvania
Based on joint research with Neil Doherty, Dwight Jaffee, Robert Meyer, Erwann Michel-Kerjan, and Mark Pauly

Howard Kunreuther began with a comment on Leary's presentation. He noted that there are important public-good aspects of adaptation: interdependence (if you fail to act, it affects your neighbors), the "ex-post issue" (if you do not adapt now, everyone else has to rescue you later), and the need to create incentive systems that give people immediate returns to overcome myopia.

[4]The presentation is available at http://www7.nationalacademies.org/hdgc/Adaptation%20to%20Climate%20Change%20Through%20Long%20Term%20Contracts.pdf [accessed September 2010].

The multi-investigator research project that led to this presentation makes three main points: (1) individuals focus on short-term horizons, and disasters are below the threshold of concern; (2) people therefore fail to take adaptation measures prior to disaster (e.g., people commonly cancel insurance after it fails to pay off in a few years, apparently thinking of it as an unproductive investment); and (3) these problems could be addressed with well-enforced long-term contracts and short-term economic incentives to deal with myopia.

Kunreuther defined the present as a new era of catastrophe. Property losses from natural hazards have been increasing over time, and insurance has failed to cover the losses—even the insured losses—so public-sector aid is needed afterward. Losses have increased much more than even the insurance industry expected. The reasons include increased urbanization and value at risk. For example, the population of Florida has increased 590 percent since 1950: the 1992 Hurricane Andrew, if it had hit in 2004, would have produced $120 billion in losses, compared with the $46 billion of losses from Hurricane Katrina. Weather patterns also have changed. More intense weather-related events, combined with sea level rise and the increased value at risk, have increased the risk significantly. Insured coastal exposure as of December 2007 was $2.5 trillion in Florida and $2.4 trillion in New York. The benefits of adaptation are therefore huge. A 500-year storm event in Florida would produce $160 billion in losses with the existing infrastructure, but having all the buildings meet building codes would save half the damage, or $82 billion.

Property owners do not invest in cost-effective adaptation measures for several reasons. One is myopia. People pay little attention and do not pay attention for long. They also underestimate the probability of costly events. Paying the up-front cost of adaptation also is a major problem (people lack the liquidity). Also, people anticipate that they will be bailed out. Many think they may be moving soon, so their investments will not pay back. Hurricanes produce a lot of damage from storm surge, and homeowners' insurance does not cover it. The National Flood Insurance Program does not cover wind damage. Thus, the insurance that people actually have does not fully cover the risks they face.

Kunreuther's group has developed a proposal for long-term flood insurance, a product that private industry has no interest in providing. He noted that, in the 1920s, mortgages normally were written for only 1-3 years, and when companies were in trouble, they refused to renew them. The private sector got into the business of offering long-term mortgages only because government started securing them. Similarly, government can offer long-term flood insurance first, and the insurance industry can get into the picture later.

The plan is to offer long-term insurance for floods and financing for adaptation, both tied to the property. The insurance rates have to reflect risk, using updated flood maps. Low-income people currently residing in flood-prone areas would be offered insurance vouchers, on a model similar to food stamps. This plan would give everyone protection and a signal for safety. It would protect homeowners from water damage from floods and hurricanes. It would encourage adaptation by giving a discount on insurance premiums for taking action. Long-term insurance is called for because people otherwise cancel their policies. In Florida, 62 percent of people who had insurance in 2000 had canceled it by 2005; in Mississippi, 83 percent no longer had it.

The program would tie the insurance to the property even after resale and would offer long-term loans for protection, with the payments becoming worthwhile because of the lower insurance premiums. Under this plan, homeowners would save money, insurers would avoid catastrophic losses, and taxpayers and the government would avoid disaster relief expenses.

The effect of climate change on long-term insurance needs analysis. With climate change the government is the only ultimate insurer against catastrophe. Data are needed on the impact of climate change on sea level rise, storm surge damage, and the effects of adaptation actions on disaster losses. Data from the United Kingdom show that adaptation combined with climate change lowers damage compared with no adaptation and no climate change. Kunreuther ended with a list of research and policy questions that need to be addressed to make choices on such things as the appropriate length of long-term contracts and ways to protect insurers against changes in risk estimates and homeowners against insolvency of insurers. Long-term flood insurance was presented as a good policy beginning for encouraging investment in adaptation; however, research is needed to determine how to incorporate climate change in such a strategy.

OPPORTUNITIES AND CONSTRAINTS TO CHARACTERIZING AND ASSESSING ADAPTIVE CAPACITY

Maria Carmen Lemos
University of Michigan

Maria Carmen Lemos observed that the literature on adaptive capacity began with a list of things that might increase the ability to adapt. Theory has become more sophisticated with time and now recognizes trade-offs and the impossibility of taking all adaptive actions at once. She noted that adaptive capacity is difficult to measure because (1) it is a latent condition (you do not know how much capacity you have until you try to use it); (2) it is dynamic and relates to time, scale, and values; (3) there is a lack of baseline data; (4) there are problems with some measurement techniques (e.g., cost-benefit analysis); (5) it may vary with scale; and (6) there are many unknowns. Moreover, adaptive capacity is nested: capacity at one scale affects capacity at other scales.

There is a need to "unpack" the concept of adaptation and related concepts (e.g., what is knowledge? technology?). Technical knowledge, such as climate model information, has equity issues. Everyone wants it (and anyone with a computer can get it), but it is used differently, access is unequal, there are dissemination constraints, and there are opportunity costs.

She noted that adaptive policies in Brazil have had varying outcomes. Of the three she examined, the one that was most apparently successful involved drought management in Ceará. In a drought, water is usually allocated for the short term, but adaptive capacity for future droughts does not increase. A two-tiered approach could make short-term adjustments, such as water distribution, combined with long-term structural reform to addresses the inequalities in vulnerability. This could be a virtuous cycle. She added that making risk management more democratic is likely to be a good strategy. She also noted that, especially in developing countries but also to some extent in the United States, adaptive capacity to climate change may be very similar to adaptive capacity generally.

In the discussion, Sanchez-Rodriguez asked if it is worthwhile to try to develop a general theory of adaptation when so much about it is specific. Lemos replied that adaptation options are greatly varied, but that adaptive capacity may be more general because it can be applied to a variety of situations. Kasperson commented that adaptive capacity is a means, not an end, so that the key questions are how much of the adaptive capacity is actually used and why. Bonnie McCay noted that adaptive strategies and response processes were a topic of general interest in the 1970s (e.g., farming practices protect against small frosts and marriage patterns protect against killing frosts). She noted the absence of reference to this literature and suggested that perhaps climate change is so catastrophic that those concepts are not applicable.

IDENTIFYING AND OVERCOMING BARRIERS TO ADAPTATION: INSIGHTS FROM THE TRENCHES OF MUDDLING THROUGH

Susanne Moser (with Julia Ekstrom)[5]

Susanne Moser explained that her presentation comes from a literature review study that looked inductively at the literature on adaptation, focus-

[5]The presentation is available at http://www7.nationalacademies.org/hdgc/Identifying%20and%20Overcoming%20Barriers%20to%20Adaptation_%20Insights%20from%20the%20Trenches%20of.pdf [accessed September 2010].

ing on barriers to adaptation—a topic that was not discussed much five years ago but has been a major topic in this workshop. It focuses mostly on planned adaptation.

Moser distinguished between limits, which are absolute thresholds, and barriers, which are things that can delay or stop adaptation processes or make them less effective and efficient. She emphasized that her framework is not normative (i.e., she does not presume that all barriers are bad and need to be overcome), but descriptive. For example, she noted that some barriers may be good to have and also that something like lack of money, which looks like a limit to someone who lacks money, may look like a barrier to a researcher. The study presented the diagnosis of barriers within a decision-making framework. While thus explicitly focused on a human system (the decision-making process, the decisions, and the decision makers), the framework does not ignore the physical or ecological constraints within which this human system exists. The framework, while actor-centric, also considers the contexts of action, including governance and the human and biophysical environments. It emphasizes processes and also is interested in outcomes. The conceptual framework is iterative: the adaptation decision-making process includes three basic phases—understanding, planning, and managing, in a circular influence diagram presented as including three phases and nine substages: understanding (problem detection, information gathering, and problem [re]definition); planning (development of options, assessment of options, selection of options); and managing (implementation, monitoring, and evaluation), all returning again to problem detection (see Figure 9-1). She noted that barriers can exist at any point in the chain but that there is little practical knowledge on the implementation (postdecision) side of the diagram, because few adaptation decisions have reached that stage.

To diagnose barriers, one must ask, at each stage of the process, what can slow the process and what causes these impediments. For example, in problem detection, the barriers can include the existence of the signal of a problem, whether people detect it, whether it passes a threshold of concern, and whether people think they can respond. For each of these kinds of barriers, there is a longer list of more specific diagnostic questions related to the various actors whose actions can help or prevent problem detection. Inductively, five barriers come up most often: leadership, resources, information and communication, participation, and cultural cognition.

How can these barriers be overcome? It depends on their type. They may be proximate or remote in terms of space and jurisdiction, and, in terms of time, they may be contemporary or result from a legacy (e.g., a law). Barriers that are remote and result from a legacy are the hardest to change.

The next steps in her research effort will be to test this model in four

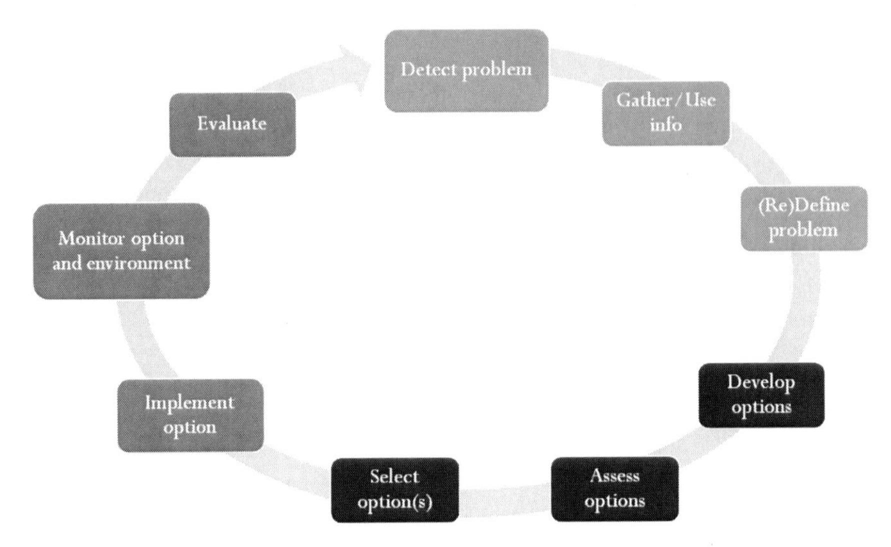

FIGURE 9-1 A schematic representation of the phases and substages of the adaptation decision-making process. The understanding phase is shown in light shading, the planning phase in dark shading, and the managing phase is in medium shading. In practical reality, decision makers may not go through each of the stages completely or in this orderly fashion.
SOURCE: Susanne Moser. Used with permission.

San Francisco Bay case studies at the municipal level, to see if the framework is useful. A larger adaptation study is also connected to this framework, with interventions being planned. A separate study will look at ways of framing the issue for different audiences.

In the discussion, Ashwini Chhatre asked about "invisible" adaptation or continuous social change. If this is happening where one is not looking for it, how can one detect its signal? He noted also that researchers are looking at adaptation only with respect to things that they perceive as needing a response. Actors get signals from everything, not just climate change. He said that looking at climate change creates a selection bias. Kasperson rephrased this idea in terms of Burton's idea of "incidental" adaptation: policy makers are asked about what was intended when they did something adaptive, but actors cannot answer questions about intention clearly for themselves. Moser agreed with the fundamental issue of barriers also arising in unplanned/incidental adaptation, but she reemphasized that this study focused on planned, deliberate, conscious adaptation processes. She also said that her study is examining actions whether or not they were undertaken explicitly as responses to climate change.

PRIORITIES FROM PRACTITIONERS FOR SOCIAL AND BEHAVIORAL RESEARCH

Roger Kasperson

Kasperson asked practitioners who were present at the workshop to suggest what they have not heard that they would like to know more about. The following topics were mentioned.

- What are some specific win-win solutions that can be pursued in adaptation contexts?
- How can the benefits of adaptation action versus business as usual best be demonstrated and quantified?
- How can incentives be aligned to favor collective action on adaptation?
- How can leadership be facilitated and the risks of leadership be reduced?
- What is the appropriate timing for infrastructure decisions, such as for actions in anticipation of future sea level rise?
- What should the federal government be doing to support and facilitate adaptation in localities, especially those with fewer resources?
- One-time studies have only a limited impact on decision makers. How can the approach be shifted to one of systematic monitoring?
- What is really meant by adaptive capacity, and how can it best be built?
- Ecosystem managers are concerned with tipping points and ecosystem collapse. What research and methods are needed to identify tipping points and to link them to stressors?
- "Mega-fauna" are an important issue in climate change. How can their importance to various stakeholders be assessed?
- The field is well short of what needs to be done in communicating about climate risks. What are the information needs and strategies for real dialogue?

When asked about important next steps, the practitioners present identified these:

- Serious efforts at capacity building
- Better coordination and collaboration and more engagement of stakeholders
- Increased dialogue to define information needs

- Caution about recommending best practices for different issue domains
- Building more inclusive knowledge networks
- Understanding that conflicts may sometimes have positive value and should not always be resisted

10

Synthesis of Key Questions for the Workshop

At the outset of the workshop, members of the organizing panel agreed to listen throughout the presentations and discussions for responses to the key questions that were identified in advance (see Box II-1). What follows are their syntheses of the responses they heard.

INITIATING ADAPTATION EFFORTS

Richard Andrews
University of North Carolina, Chapel Hill

Two key questions were posed to the participants about initiating adaptation efforts:

1. What are the main barriers to initiating adaptation efforts, and what has been effective at overcoming them?
2. How and under what conditions have climate change considerations been successfully integrated into the normal activities of regional or sectoral risk management organizations?

Andrews heard workshop participants identify several kinds of barriers that can prevent initiating adaptation efforts. These include barriers internal to each decision maker and to other actors (e.g., attitudes, incorrect knowledge) and barriers external to them (e.g., lack of resources, lack of authority or jurisdiction to act, misaligned incentives). Barriers can be found at the levels of individuals, of organizations, and of communities, as well as at higher levels of regional to global governance. Although it may be tempting to identify such barriers as "maladaptations," more often they represent the legacy of adaptations that were successful in the past. Individuals who now benefit from them are therefore reluctant to give them up. At the organizational and community levels, barriers include inertia and aversion to change, low public awareness, old battle lines, traditional organizational missions, and organizational "silos," among others. Misaligned incentives are important but are not the only important barriers.

He also heard several creative ideas and recurrent themes about ways of overcoming these barriers. Some lessons can be learned from analogous fields, such as public health.

One key step to overcome barriers is to integrate the responsibility for considering potential climate change impacts and proposing alternatives for adaptation into the core missions and operating decision processes of all organizations that might be affected. Doing this is essential in order to identify for decision makers (and the public) how adaptation to anticipated climate change is important to their core responsibilities and how it might necessitate different choices than would have been made in the past. This process of regularizing attention to the adaptation issue is similar to the process pioneered by the environmental impact statement requirement of the National Environmental Policy Act 40 years ago.

A second theme is that, in addition to integrating climate change into routine organizational decision processes, there may also be the need for a high-level convening entity with the responsibility to address interagency and interjurisdictional conflicts and coordination needs.

A third recurrent theme is that empowering bottom-up initiatives and promoting peer-to-peer learning among them is a key to more widespread commitment to adaptation initiatives. Decision makers in cities, communities, and businesses are far more likely to trust and adopt the approaches of successful peers facing similar issues, who can speak to the need, the benefits, and methods of implementation, than to trust officials at higher levels of government or academics. The group heard repeatedly about the importance of empowering bottom-up responses, for example by cities, as well as by other localized and regional networks of stakeholders and scientists—for example, initiatives of the Regional Integrated Sciences and Assessments (RISA) Program of the National Oceanic and Atmospheric Administration (NOAA).

Fourth, windows of opportunity are important in framing proposals for change and overcoming barriers, and it is important to be prepared to use such opportunities effectively when they arise. Localized or isolated disasters may be leading indicators of more serious potential impacts and patterns of impacts that may occur in the future if preventive adaptation does not occur. A central argument for preventive adaptation is to diminish the need for highly expensive postdisaster response and persistent disruptive changes.

Fifth, there is a need to reframe climate change adaptation as opportunities to protect and achieve commonly shared values for the future of a community rather than as threats to entrenched values and interests. There is also a need to bring new actors into the process who can help infuse new ways of thinking and generate creative solutions.

In answer to the second core question, one key factor for success in adaptation initiatives is buy-in by high-level local leadership, often resulting from peer-to-peer learning among such leaders. A second is promoting interaction among technical and stakeholder organizations to integrate the best knowledge that is available both from experts and from the communities affected by climate change and adaptation measures. A critical third factor is continuous engagement with adaptive communities rather than one-time studies, rooting adaptation in local areas and recognizing and using relevant local knowledge, which may be as important, or more important, than downscaled global models and other more general expertise. A key value of social science in these processes is to help all participants visualize options and consider their costs and consequences. Finally, it is important to start by focusing on a few immediate decision issues that link to community priorities and that can be acted on immediately with available information, for which the consideration of climate change could make a significant difference to the outcome.

Finally, Andrews identified a number of unanswered questions:

- How can action be induced among cities and organizations that are not yet active?
- How does one deal with equity questions and with highly vulnerable populations—both by engaging them in decisions and by addressing their interests and needs?
- How can long-term adaptive institutions be created, such as long-term insurance contracts?
- How can people get default options right, to facilitate both organizational and individual behavior patterns that are most consistent with effective adaptation?

In the discussion that followed, Cynthia Rosenzweig noted that having an ongoing series of foundational reports, such as the U.S. National Assessment of Climate Change, can help bring information into the system. Kristie Ebi said that a lot of work needs to be done in the scientific community, which does not understand risk management. For example, in general, Working Group I of the Intergovernmental Panel on Climate Change still says to wait for more certainty and does not understand what Working Group II needs.

JoAnn Carmin commented that much of the discussion wrestles with the problem of diffusion but noted that adaptation also requires innovation before diffusion can happen.

Maria Blair said that the questions are great, but there is nothing to manage yet. There was division in the room about whether anything is known about how to make adaptation happen. Questions were asked about what is being managed and why. Adaptation is happening, and, in the United Kingdom, people are even thinking about ways to adapt to a world that is 4 degrees Celsius warmer than before. Such specificity is not evident in this country, where policies have rarely designated an explicit end point to manage to or for. She questioned whether, if something like a 4 degree temperature increase is anticipated, "managing" is even a realistic idea. What to do with that kind of a transformation is not addressed. Some literatures talk about "navigating" transitions. She noted that there is a history of management approaches that prop up maladaptive systems. People assume that they can do better in the future, but is there any evidence? Maybe it would be better to focus more on getting money, technology, and information to the people who will be adapting. She identified the adaptation problem as a collective action problem, involving more than just choices by individual actors, and suggested that institutional change is therefore a way to proceed. She had not heard in the workshop about how to manage institutions. There may be much better thinking about short-term than longer term adaptations, but people need to get over short-term thinking, she said. It is not clear which short-term adaptations will be maladaptive in the long term. For example, if there is dramatic, abrupt sea level rise, almost everything now being considered is maladaptive.

COORDINATING ADAPTATION EFFORTS

Stewart Cohen
Environment Canada and University of British Columbia

Three key questions were posed to the participants about coordinating adaptation efforts:

1. What strategies or methods have been effective for coordinating adaptation efforts across scales (e.g., national, regional, local, individual)?
2. What strategies or methods have been effective for coordinating adaptation efforts across sectors (e.g., government, private, nonprofit, community)?

3. How should stakeholders and the public be engaged in adaptation efforts?

Cohen began by noting a set of coordination issues that had been raised at the workshop: creating common space; building bridges and/or webs; the issue of what to adapt to and who will be doing it (future climate change is different from past variability, and coordination will have to include new actors); the need for champions of coordination to create dialogue (e.g., among government agencies, trade associations, specific government programs, such as RISA, collaborations among cities); a need to integrate climate change adaptation with long-established hazards and disaster research networks; and a need for effective flow of climate change information to stakeholders.

Cohen said that a participatory approach is important and implies a complicated set of links, which involve a variety of translators, many of whom are practitioners who use their tools of practice to inform the stakeholders they work for. The tool developers themselves may or may not be good translators.

He suggested that group-based model building could be a useful option to combine technical information and experiential knowledge. This may be more attractive than having people experiment with a tool that someone else has built. However, it needs a platform to make it possible, and the feasibility of the approach will vary depending on the backgrounds and skills of participants.

He noted that a lot was said at the workshop about the role of networks in building webs. Building webs requires a two-way process, not one-way outreach. Finding the right partners can be a problem. In some networks, as suggested by NOAA's RISA centers, experts can become extension specialists for local adaptation. The language of networks is helpful, even for people who do not think about the morphology of network structures, because it justifies the roles of coordination enablers. Who will be an enabler of change is not known in advance, but the community may know who can fill that role.

Cohen noted that "mainstreaming" is a means objective, not an ends objective. Dangers in mainstreaming include the potential to disempower people and to give too much attention to sources of funding. He also said that coordination may hide underlying sources of conflict. It can be useful to turn a conflict into an argument, in which people can learn from each other. Coordination may help by bringing different actors into the room.

How can coordination improve response capacity? One way is through shared learning (aimed at solving a problem, using local knowledge, and translating information from unfamiliar sources). Another way is to build institutional memory of information exchanges, which does not automatically happen with one-time projects. A third way is through dialogue in adaptation research efforts, which can go beyond outreach and opinion surveys, especially if the dialogue is facilitated by local partners. This can include both group-based model building and model-enabled dialogue.

In the discussion that followed, Ebi noted that there are some strong adaptation networks internationally that are developing institutional memory. Rosenzweig said local practitioners cannot be counted on to remain in place, so there is a need to leave something behind for local organizations.

INFORMING ADAPTATION EFFORTS

Michele Betsill
Colorado State University

Two key questions were posed to the participants about informing adaptation efforts:

1. What methods have been successful in providing needed information to risk managers who must cope with climate change?
2. How should efforts to inform climate adaptation characterize risk and uncertainty about future climate and other processes affecting climate risk?

Regarding successful methods, Betsill concluded that the key is process, not products. However, the process needs to be organized around a task or problem; has to be sustained in order to enable learning, build trust, and allow for updating of information; and has to involve the users of information in partnership with researchers and agency officials. Information production has to respond to users' needs, with coproduction of knowledge. Sometimes the scientific credibility of information is less important than issues of trust and salience.

A second lesson is that one size does not fit all. Information products are needed, and many types of information need to be presented in many ways. Regarding ways to characterize risk and uncertainty, Betsill said she had not heard much about this. She noted that people do know how to deal with uncertainty. The workshop heard about the importance of providing choices and empowering people—giving them a range of things to do and ways to think about the choices.

In the discussion, Kirstin Dow noted some new dimensions in informing adaptation decisions. There are new forms of technology, such as new social media. It is possible, using the new media, to talk with people in multiple locations to take advantage of connectivity among local places.

Several participants engaged in a discussion of the characterization of uncertainty in informing adaptation. Michael Savonis said that climate scientists are very conservative about drawing conclusions and, from a policy maker's perspective, he preferred that they convey both the uncertainty and what they do know. He realizes that there is a very strong element of judgment, but decision makers rely on scientific input that uses judgment because of what lies behind it. Scientists need to explain why they believe what they believe. Decision makers want more than just "what the model says." They also want scientists' evaluations of what the model says. Another participant noted, however, that modelers are often not the same people who collected the data, so they may not know how to evaluate the strength of evidence for certain model outputs.

Cohen commented that there are uncertainties both in models and in the translation of models for decision makers. Models need to be translated carefully. For example, timber supply is not the same as the biomass of the wood, and water supply is not the same as stream flow. The challenge is to connect model results to decision needs—to translate them into terms the decision maker understands. That translation process is currently poorly understood.

Dow commented that some decision makers she knows actually want numerical probabilities. Amy Luers agreed, noting that with climate change, uncertainty has different forms: what can be influenced (e.g., emissions) and what cannot (e.g., climate sensitivity).

SCIENCE NEEDS FOR ADAPTATION EFFORTS

Maria Carmen Lemos

Three key questions were posed about science needs for informing adaptation efforts:

1. What new social science knowledge is needed to develop a national adaptation strategy?
2. What metrics and indicators are needed to support adaptation decisions (e.g., indicators of vulnerability, resilience, adaptive potential, effectiveness of adaptation efforts)?
3. What are the key needs for databases, scenario development, and modeling?

Lemos noted that there have been many efforts to develop science priorities for human dimensions of climate change, including at least three recent National Research Council reports. They all agree that more social science is needed. This workshop is an opportunity to identify which items on the researchers' lists correspond to what decision makers say they need.

The way the decision makers phrase their needs is illuminating. One issue is how to make decision makers care, given such barriers to action as social inertia and the complexity of government. Related to this is the question of how science can be used to mobilize people. Lemos heard the government officials saying that decision makers care about thresholds and tipping points, potential catastrophe, security and national interest, the linkage between climate change, as well as things of concern on a day-to-day basis, win-win situations (e.g., responses that create jobs and wealth). They said that decision makers care about setting priorities for their investments. Even if the concept of adaptive capacity is not ideal, it does help to decide where to invest. They also care about questions of when people need to know what and of how to learn from experience. Finally, there was an expressed need during the workshop for scientists to communicate what they know better: synthesizing it to make it easy for decision makers to use it.

The scientists' agenda represents only the group that is present, but there are some frequently repeated terms: the roles of values, beliefs, and attitudes in action and inaction; networks; the roles of formal and informal institutions, those that like to change or resist change, and effects of rules on action; surveying frameworks for thinking about adaptation and carrying out metastudies; understanding the evolution of preferences; social inertia, trade-offs, the role of management and governance, the organization of adaptation options, and ways to evaluate choices. The big themes on the scientists' research agenda are how to understand system change from a social and ecological point of view and how to apply knowledge of social change to climate.

In the discussion, Roger Kasperson characterized these comments as identifying important questions about which quite a bit is known. With sufficient resources, someone could pull this knowledge out of the research literature to make it useful to decision makers. The research community cannot give pithy answers to complex questions, but it can identify important topic areas that can provide something to the decision makers and it can tell them where to find out more. Lemos added that the research community could grow very quickly if it attracts researchers who know about poverty, preferences, and other fundamental social science topics and get them thinking about climate issues.

Kasperson commented that the idea of good communication as two-way was settled in the social science literature 30 years ago, but that, except for the RISA Program, it is still not understood in practice. Instead, many people talk about outreach, rather than about inreach, and get nowhere. Conversations about informing are still centered on dissemination and influence. Susanne Moser said that both government officials and scientists are trained just to get their messages out and suggested that part of the problem is that students have not been properly educated. Another participant cited the work of Ralph Keeney and others on value-tree analysis, a technique that gets values on the table as a basis for discussion and for increasing understanding and that works by forcing two-way communication. Helen Ingram noted that issues get on the agenda if there are scheduled decisions to make and if they are linked with social mobilization. She added that researchers do not normally write to political imperatives.

Neil Leary added that communication has worked fairly well when there is a common task that requires iteration. In contrast, climate scientists and risk research scientists talk to each other about what they want from each other, and progress is not made. However, if such a discussion is part of a process with a goal and continues for a while, they can figure it out.

Ashwini Chhatre said that collaboration can go too far, in the sense that legitimizing research only in terms of what decision makers want can lead to failure to do the basic research that is needed. Jamie Kruse said that one science need of importance to NOAA is to have good performance measures for adaptation.

MANAGING ADAPTATION EFFORTS

Susanne Moser

Three key questions were posed about managing adaptation efforts:

1. How should a national climate adaptation effort set priorities across hazards, sectors, regions, and time? What criteria, and what processes, should be used?
2. What mechanisms can help make adaptation efforts adaptive? How can a system enable decision makers to learn efficiently from experience?
3. What additional capacity do federal agencies need to support adaptation and resilience?

Moser began by recalling Blair's comment that it may be premature to say much about how to "manage adaptation" because the nation is still at such an early stage in the adaptation process. Very few projects, communities, or states have actually gotten to the point of implementing their plans or making actual changes on the ground.

She also noted that perhaps it is difficult to decide where to focus a long-lasting, deliberately learning-oriented, iterative process, because decision makers usually are distracted by having to deal with the crisis of the day. Moreover, so many things are critical—water, coastal areas, food security, species protection—that both scientists and decision makers are hard pressed to say that one is more important than another. She pointed out, recalling Neil Adger's presentation, that the criteria to use for priority setting will be different depending on goals (e.g., reducing vulnerability versus efficient adaptation versus getting to system resilience). In the United States, she noted, people seem to be concerned mainly with efficient adaptation and are rarely concerned with the original meaning of resilience or with reducing differential vulnerability. She challenged Blair and Kathy Jacobs (who had emphasized different goals in their respective areas of work) to clarify for and among themselves (i.e., for the federal government) what they actually want to facilitate and support.

Regarding ways to make adaptation more adaptive, Moser noted that the process needs to be deliberative, iterative, and well facilitated, with feedback and dense networks. But commitment, institutionalization, and leadership will also be necessary to make adaptation an ongoing consideration. Whereas it may have been possible to deal with problems once and for all in the past, a continually changing climate does not allow this luxury. She also noted that the workshop heard about the benefit of conflict as an opportunity to revisit issues. One way to institutionalize such ongoing processes is to change the expectations for professional performance that are embedded in job descriptions: from expectations for perfect, flawless, or successful outcomes to expectations for learning from past decisions. Accountability is one of the quickest ways to ensure that learning happens.

On capacity needs, Moser pointed to Lemos's list of components of adaptive capacity: human capital (educating the current and future generations); information and technological resources; material resources and infrastructure (critical for functioning but possibly not necessary for adaptation); social capital (e.g., more democratic forms of engagement in the adaptation process and trust, which is slow to build and quick to lose); political capital (the existence of visible leaders, which makes it easier for others to act); wealth and financial resources; and public–private partnerships as a mechanism. Moser noted that the experience from some of the early actors suggests that not all of these capacities are always necessary. Leaders and early adopters make do with what they have and position themselves for a longer term process that allows them to build capacities they do not currently have at a later time. She also recalled examples of institutions and entitlements that have created capacity but also gave people investments in the status quo that functioned as barriers to change rather than facilitating it.

In the discussion that followed, a participant noted that, although there has been a lot of talk about adaptation planning, it is evaluation that is really important. Resources are needed for evaluation. Although this is an issue in many other environmental management and change processes, it may be even more critical in the context of adaptation to a changing climate. The management tools also need to include professional standards and norms and regulatory requirements. Another participant added that more work is needed on measurement of adaptation success.

Rosenzweig noted the importance of a conceptual frame that covers the many adaptive actions involved in managing a complex set of responses. For example, the adaptation process in New York City builds on London's idea of flexible adaptation pathways, using a diagram showing acceptable risks and how climate change is leading beyond what is acceptable. This diagram is an important management tool both to convey the need and urgency of adaptation and to trigger changes in policy as certain thresholds are reached. Moreover, such a diagram can show, in a simple model, what mitigation and adaptation do and how they complement each other. A term like "climate-resilient cities"—a common and galvanizing language—is also important to facilitate action.

COMMENTS ON MAJOR INSIGHTS AND ISSUES

Thomas Dietz

Dietz commented that behavioral and social science work on climate change adaptation needs to be in Pasteur's quadrant (Stokes, 1997). It needs to contribute to fundamental understanding, and it needs to be useful. The field needs to keep addressing the fundamental questions that are special to this area. For example, there are normative questions about what people might lose as a result of climate change. Research on adaptation can develop understanding about what people want to preserve and do not want to lose. Also, this area can be an important test bed for theories of social change.

Another big issue is risk. It is important to clarify the risks of climate change and bring in literature on how to deliberate about risk. Research needs to keep in mind the use of uncertainty as a political weapon. It is essential to know about how to identify and engage stakeholders and how to inform the codesign of processes to inform adaptation. Engagements in climate change adaptation should be treated as experiments and evaluated. People should be skeptical of claims of program success. Dietz also noted that new technologies will transform society and could also help to develop better data.

There are also important issues about research methods, he said. A lot of good case-based studies were mentioned at the workshop. In addition, more comparative work and more meta-analysis are needed. A knowledge base that derives from a multimethod, cumulative literature is not yet a reality.

Funding is an important need. Social science funding has declined over the 20 years of the U.S. Global Change Research Program, and this needs to be reversed. There is no routine forum like this workshop, in which social scientists can talk to each other in depth to advance the science. This kind of problem-oriented forum is important because the social science disciplines tend to retard the process.

Dietz pointed to the need to connect to the coupled natural and human systems work, such as is being supported by the National Science Foundation, and to human ecology and other disciplines that go beyond the social sciences. He said that networks are central, also pointing to a rich set of analytical tools that can be used as practical and theory-building tools.

CONCLUDING DISCUSSION

Kasperson invited each participant to identify one thing that should not be forgotten in the report of the workshop.

Several comments addressed the needs for improving the theory and general understanding of adaptation. Adam Henry said it is very hard to make useful recommendations without a comprehensive theory of adaptation. What there are so far are some major categories of variables, not a theory. Kasperson agreed that integrative theory is needed in order to manage and design experiments. Rosenzweig said that it is time to set up national and regional coordinated long-term efforts on adaptation, with evaluation built in, to help develop the theory. Dietz said that the commons literature shows how a research community, working within a theoretical framework, can move understanding forward.

Other comments focused on particular scientific issues. One participant noted that the idea of social inertia is important but overly simplistic and needs to be unpacked. One pointed to the need for more discussions of the critical roles of boundary organizations. One said that adaptation strategies, even in urban areas, should be analyzed within their ecosystems. One identified a need to talk about uncertainties in many knowledge areas, not only in climate science. There were also suggestions to consider information technology as a force shaping social change and to find balance between "thin" and "thick" descriptions of adaptation processes.

Several comments focused on issues of practice. One participant reiterated the importance of integrating adaptation into the workings of existing institutions. However, another noted that there is a tension between normalizing (mainstreaming) adaptation and not normalizing it. A participant noted the importance and unavoidability of conflict and said that the way to address it was not to stifle it but to use it to surface unheard voices and unrecognized impacts. Another pointed out the importance of continued "care and feeding" of local adaptation efforts after research projects end.

Some comments focused on the magnitude of the adaptation challenge. Moser called for an end to the idea that climate change will be a slow, gradual process and for an increased focus on the big adaptation challenges society may be facing. She predicted that it will be a much bigger challenge than people think. Peter Banks suggested that people are facing three kinds of adaptation simultaneously: a wrenching energy adaptation over the next 50 years, the addition of additional 180 million people in the United States, and climate change. He suggested that this combination of adaptations will lead to social turbulence and a rethinking of institutions.

A few comments emphasized the connections between science and practice. One pointed to an urgent need for social science research as a basis for practice, rather than the consulting input that provides the conventional wisdom. Others wanted more practitioners in discussions like this one, including local practitioners, to talk with scientists to build understanding and new ways of thinking. One said there is a need to follow up meetings like this one, for deeper engagement between scientists and practitioners. Another called for multidisciplinary funding of rich environments that reach from theory to practice.

Some comments focused on agencies' needs. One agency participant said that federal agencies are starting to think hard about adaptation issues, are really interested in what the research community has to say, and will be persistently asking questions of the research community. Another asked for advice on how best to set up national, state, and local networks.

Finally, several comments raised capacity issues. One participant noted the declining budget for human dimensions research and the need to develop activities and capacity in developing countries. Another noted the need to nurture the next generation of people to fill the gap between science and its application to adaptation. Bridging the gap is a job people can have. Another emphasized the need for universities to train practitioners and "boundary people" through degree programs. Dietz commented that extension services, which provide boundary people, are being gutted at the state level.

References

Abrahamson, V., Wolf, J., Lorenzoni, I., Fenn, B., Kovats, S., Wilkinson, P., Adger, W.N., and Raine, R. (2009). Perceptions of heat-wave risks to health: Interview-based study of older people in London and Norwich, UK. *Journal of Public Health, 31*(1), 119-126.

Bazerman, M.H., and Watkins, M.D. (2004). *Predictable Surprises—The Disasters You Should Have Seen Coming and How to Prevent Them.* Boston: Harvard Business School Press.

Bostrom, A., Morgan, M.G., Fischhoff, B., and Read, D. (1994).What do people know about global change? 1. Mental models. *Risk Analysis: An International Journal, 14*, 375-387.

Burt, R.S. (2004). Structural holes and good ideas. *American Journal of Sociology, 110*, 349-399.

Carson, R. (1962). *Silent Spring.* Boston, MA: Houghton Mifflin.

Dietz, T., Gardner, G.T., Gilligan, J., Stern, P.C., and Vandenbergh, M.P. (2009). Household actions can provide a behavioral wedge to rapidly reduce U.S. carbon emissions. *Proceedings of the National Academy of Sciences, 106*, 18,452-18,456.

Douglas, M., and Wildavsky, A. (1982). *Risk and Culture: An Essay on the Selection of Technological and Environmental Dangers.* Berkeley: University of California Press.

Dunlap, R.E., and McCright, A. (2008). A widening gap: Republican and Democratic views on climate change. *Environment, 50*(5), 26-35.

Dunlap, R.E., and McCright, A. (2010). Climate change denial: Sources, actors, and strategies. In C. Lever-Tracy (Ed.), *Routledge Handbook of Climate Change and Society* (Chapter 14, Part IV: Social Recognition of Climate Change). New York: Routledge.

Ebi, K.L., Smith, J., and Burton, I. (Eds.). (2005). *Integration of Public Health with Adaptation to Climate Change.* Oxford, England: Taylor & Francis.

Ehrhardt-Martinez, K., Donnelly, K.A., and Laitner, J.A. (2010). *Advanced Metering Initiatives and Residential Feedback Programs: A Meta-Review for Household Electricity-Saving Opportunities.* Washington, DC: American Council for an Energy-Efficient Economy.

Farber, D.A. (2007). Basic compensation for victims of climate change. *University of Pennsylvania Law Review, 155*, 1,605-1,656. Available: http://www.pennumbra.com/issues/pdfs/155-6/Farber.pdf [accessed September 2010].

Fischhoff, B. (1982). Debiasing. In D. Kahneman, P. Slovic, and A. Tversky, Eds. *Judgment Under Uncertainty: Heuristics and Biases* (Part VIII: Corrective Procedures, Chapter 31, pp. 422-445). Cambridge, England: Cambridge University Press.

Fleishman, L., Bruine de Bruin, W., and Morgan, M.G. (2010). Informed public preferences for electricity portfolios with CCS and other low-carbon technologies. *Risk Analysis: An International Journal, 30*(9), 1,399-1,410.

Gardner, G.T., and Stern, P.C. (2002). *Environmental Problems and Human Behavior, Second Edition.* Boston: Pearson.

Gardner, G.T., and Stern, P.C. (2008). The short list: The most effective actions U.S. households can take to curb climate change. *Environment, 50*(5), 12-21.

Granade, H.C., Creyts, J., Derkach, A., Farese, P., Nyquist, S., and Ostrowski, K. (2009). *Unlocking Energy Efficiency in the U.S. Economy.* McKinsey Global Energy and Materials. Available: http://www.mckinsey.com/clientservice/electricpowernaturalgas/downloads/US_energy_efficiency_full_report.pdf [accessed August 2010].

Jacques, P.J., Dunlap, R.E., and Freeman, M. (2008). The organization of denial: Conservative think tanks and environmental skepticism. *Environmental Politics, 17*(3), 349-385.

Leiserowitz, A., Maibach, E., and Roser-Renouf, C. (2008). *Global Warming's "Six Americas": An Audience Segmentation.* Yale University and George Mason University. New Haven, CT: Yale Project on Climate Change.

Leiserowitz, A., Maibach, E., and Roser-Renouf, C. (2009). *Climate Change in the American Mind: Americans' Climate Change Beliefs, Attitudes, Policy Preferences, and Actions.* Yale University and George Mason University. New Haven, CT: Yale Project on Climate Change.

Michaels, D. (2008). *Doubt Is Their Product: How Industry's Assault on Science Threatens Your Health.* New York: Oxford University Press.

Mileti, D. (1999). *Disasters by Design: A Reassessment of Natural Hazards in the United States.* Washington, DC: Joseph Henry Press.

Moser, S.C., and Ekstrom, J.A. (2010). Barriers to climate change adaptation: A diagnostic framework. Submitted to *Proceedings of the National Academy of* Sciences.

National Research Council. (1985). *Energy Efficiency in Buildings: Behavioral Issues.* Committee on Behavioral and Social Aspects of Energy Consumption and production. P.C. Stern, Ed. Washington, DC: National Academy Press.

National Research Council. (1996). *Understanding Risk: Informing Decisions in a Democratic Society.* P.C. Stern and H.V. Fineberg, Eds. Washington, DC: National Academy Press.

National Research Council. (2008). *Research and Networks for Decision Support in the NOAA Sectoral Applications Research Program.* Panel on Design Issues for the NOAA Sectoral Applications Research Program. H.M. Ingram and P.C. Stern, Eds. Committee on the Human Dimensions of Global Change, Division of Behavioral and Social Sciences and Education. Washington, DC: The National Academies Press.

National Research Council. (2009). *America's Energy Future.* Committee on America's Energy Future. Washington, DC: The National Academies Press.

National Research Council. (2010a). *Adapting to the Impacts of Climate Change.* Panel on Adapting to the Impacts of Climate Change. Board on Atmospheric Sciences and Climate. Division on Earth and Life Studies. Washington, DC: The National Academies Press.

National Research Council. (2010b). *Advancing the Science of Climate Change.* Panel on Advancing the Science of Climate Change. Board on Atmospheric Sciences and Climate. Division on Earth and Life Studies. Washington, DC: The National Academies Press.

National Research Council. (2010c). *Limiting the Magnitude of Future Climate Change.* Panel on Limiting the Magnitude of Future Climate Change. Board on Atmospheric Sciences and Climate. Division on Earth and Life Studies. Washington, DC: The National Academies Press.

Nolan, J., Schultz, P.W., Cialdini, R.B., Griskevicius, V., and Goldstein, N. (2008). Normative social influence is underdetected. *Personality and Social Psychology Bulletin, 34*, 913-923.

Pacala, S., and Socolow, R.H. (2004). Stabilization wedges: Solving the climate problem for the next 50 years with current technologies. *Science, 305*, 968-972.

Palmgren, C.R., Morgan, M.G., Bruin, W.B.d., and Keith, D.W. (2004). Initial public perceptions of deep geological and oceanic disposal of carbon dioxide. *Environmental Science & Technology, 38*, 6,441-6,450.

Parry, M.L., Canziani, O.F., Palutikof, J.P., van der Linden, P.J., and Hanson, C.E. (Eds.). (2007). *Impacts, Adaptation, and Vulnerability. Contribution of Working Group II to the Fourth Assessment Report of the Intergovernmental Panel on Climate Change.* Cambridge, England: Cambridge University Press.

Pidgeon, N., Kasperson, R., and Slovic, P. (Eds.). (2003). *The Social Amplification of Risk.* Cambridge, England: Cambridge University Press.

Pidgeon, N.F., Poortinga, W., Rowe, G., Horlick-Jones, T., Walls, J., and O'Riodan, T. (2005). Using surveys in public participation processes for risk decision making: The case of the 2003 British GM Nation? Public debate. *Risk Analysis: An International Journal, 25*, 467-479.

Pidgeon, N.F., Lorenzoni, I. and Poortinga, W. (2008). Climate change or nuclear power—no thanks! A quantitative study of public perceptions and risk framing in Britain. *Global Environmental Change, 18*, 69-85.

Pidgeon, N.F., Henwood, K.L., Parkhill, K.A., Venables, D., and Simmons, P. (2008). *Living with Nuclear Power in Britain: A Mixed-Methods Study. Summary Findings Report.* Cardiff University and the University of East Anglia. Available: http://www.kent.ac.uk/scarr/SCARRNuclearReportPidgeonetalFINAL3.pdf [accessed September 2010].

Pidgeon, N.F., Harthorn, B., Bryant, K., and Rogers-Hayden, T. (2009). Deliberating the risks of nanotechnology for energy and health applications in the U.S. and U.K. *Nature Nanotechnology, 4*, 95-98.

Read, D., Bostrom, A., Morgan, M.G., Fischhoff, B., and Smuts, T. (1994). What do people know about global change? 2. Survey studies of educated laypepople. *Risk Analysis: An International Journal, 14*, 971-982.

Reynolds, T.W., Bostrom, A., Read, D., and Morgan, M.G. (2010). Now what do people know about global climate change? Survey studies of educated laypeople. Submitted to *Risk Analysis: An International Journal.*

Sabatier, P.A., and Jenkins-Smith, H.C. (Eds.). (1993). *Policy Change and Learning: An Advocacy Coalition Approach.* Boulder, CO: Westview.

Schneider, M., Scholz, J., Lubell, M., Mindruta, D., and Edwardsen, M. (2003). Building consensual institutions: Networks and the National Estuary Program. *American Journal of Political Science, 47*, 143-158.

Schultz, P.W., Nolan, J., Cialdini, R.B., Goldstein, N., and Griskevicius, V. (2007). The constructive, destructive, and reconstructive power of social norms. *Psychological Science, 18*, 429-434.

Solomon, S., Plattner, G.K., Knutti, R., and Friedlingstein, P. (2009). Irreversible climate change due to carbon dioxide emissions. *Proceedings of the National Academy of Sciences, 106*(6), 1704-1709.

Spence, A. and Pidgeon, N.F. (2009). Psychology, climate change, and sustainable behavior. *Environment: Science and Policy for Sustainable Development, 51*(6), 8-18.

Spence, A., Poortinga, W., Pidgeon, N.F., and Lorenzoni, I. (2010). Public perceptions of energy choices: The influence of beliefs about climate change and the environment. *Energy and Environment, 21*(5), 385-407.

Stern, P.C. (1986). Blind spots in policy analysis: What economics doesn't say about energy use. *Journal of Policy Analysis and Management, 5*, 200-227.

Stern, P.C., Gardner, G.T., Vandenbergh, M.P., Dietz, T., and Gilligan, J. (2010). Design principles for carbon emissions reduction programs. *Environmental Science & Technology, 44*, 4,847-4,848.

Stokes, D.E. (1997). *Pasteur's Quadrant: Basic Science and Innovation*. Washington, DC: Brookings Institution Press.

Swim, J., Clayton, S., Doherty, T., Gifford, R., Howard, G., Reser, J., Stern, P., and Weber, E. (2009). *Psychology and Global Climate Change: Addressing a Multifaceted Phenomenon and Set of Challenges. A Report of the American Psychological Association Task Force on the Interface Between Psychology and Global Climate Change*. Washington, DC: American Psychological Association.

Thaler, R.H., and Sunstein, C.R. (2008). *Nudge: Improving Decisions About Health, Wealth, and Happiness*. New Haven, CT: Yale University Press.

U.S. Climate Change Science Program. (2008). *Decision-Support Experiments and Evaluations Using Seasonal-to-Interannual Forecasts and Observational Data: A Focus on Water Resources*. Final Report Synthesis and Assessment Product 5.3. N. Beller-Simms, H. Ingram, D. Feldman, N. Mantua, K.L. Jacobs, and A.M. Waple, Eds. Washington, DC: Author.

U.S. Energy Information Agency. (2008). *Emissions of Greenhouse Gases in the United States 2007*. Pub. No DOE/EIA-0573, Table 5. Washington, DC: U.S. Department of Energy.

U.S. Global Change Research Program. (2009a). *Climate Literacy: The Essential Principles of Climate Sciences, A Guide for Individuals and Communities*. Washington DC: Author. Available: http://www.climatescience.gov/Library/Literacy/ [accessed September 2010].

U.S. Global Change Research Program. (2009b). *Global Climate Change Impacts in the United States*. New York: Cambridge University Press.

Vandenbergh, M.P., Stern, P.C., Gardner, G.T., Dietz, T., and Gilligan, J. (2010). Implementing the behavioral wedge: Designing and adopting effective carbon emissions reduction programs. *Environmental Law Reporter, 40*, 10,545-10,552.

Wolf, J., Adger, W.N., Lorenzoni, I., Abrahamson, V. and Raine, R. (2010). Social capital, individual responses to heat waves and climate change adaptation: An empirical study of two UK cities. *Global Environmental Change, 20*(1), 44-52.

Appendix A

December 2009 Workshop Agenda and List of Participants

Workshop on Issues in Public Understanding and Mitigation of Climate Change
Agenda and List of Participants
December 3-4, 2009

This workshop, the first of two sponsored at the National Academies by the William and Flora Hewlett Foundation, will include four half-day sessions devoted to the following topics of pressing interest:

- *Public Understanding of Climate Change*
- *Opportunities for Limiting Climate Change Through Household Action*
- *Public Acceptance of Energy Technologies*
- *Organizational Change and the Greening of Business*

Each session will begin with presentations of current knowledge by leading social and behavioral researchers and will proceed to discussions of the practical implications of the knowledge for action by governmental and nongovernmental organizations tasked with responding to climate change. It is hoped that the discussions will stimulate participants to undertake future activities, such as new policies, programs, or research activities, to develop and implement insights arising from the workshop.

Session #1—December 3, 2009
Public Understanding of Climate Change

Climate change as a phenomenon has attributes that make it is extremely difficult for nonspecialists to understand. For example, although people typically rely on their senses and personal experience to assess conditions in the external environment, these sources are a poor guide to whether the global climate is changing or to the effects of such change. People often apply cognitive short-cuts to make sense of complex topics, but doing this with climate change easily promotes misunderstanding. The short-cut of relying on trusted sources of information is problematic because conflicting information sources claim expertise on climate change. The polarization of U.S. public opinion on climate change can be traced to such social and psychological processes.

This session will present the current state of knowledge about how nonspecialists attempt to comprehend climate change and why public opinion has become increasingly polarized, even as scientific opinion has become less so. It will conclude with discussion of what might be done about this situation—in education, in the mass media, and through the communication efforts of the nation's scientific community.

Welcoming comments, Roger Kasperson, Clark University, Panel Chair
Anthony Leiserowitz, Yale University, Session Moderator

Presentations:
Why is climate change hard to understand?—Susanne Moser, Susanne Moser Research and Consulting
Mental models of climate change—Daniel Read, Yale University
Insights from research on risk perception—Elke Weber, Columbia University
The polarization of public opinion—Riley Dunlap, Oklahoma State University

Comment and discussion topics:
—Implications for climate change education
—Implications for the mass media
—Implications for scientific communication

Discussants:
Frank Niepold, Climate Program Office, National Oceanic and Atmospheric Administration
Bud Ward, Yale Forum on Climate Change and the Media

Session #2—December 3, 2009
Opportunities for Limiting Climate Change Through Household Action

The most commonly proposed strategies for limiting climate change—developing low-carbon energy technologies and creating systems that put a price on greenhouse gas emissions—are likely to take a decade or more to yield appreciable reductions. Changing the adoption and use of existing technology can yield savings much faster if the requisite behavioral changes can be brought about.

This session will focus on the potential in the household sector—direct energy use in homes and nonbusiness travel—which accounts for about 38 percent of U.S. energy use. It will present new estimates of the technical and reasonably achievable potential in this sector and knowledge about the effectiveness of various strategies for achieving this potential. It will conclude with discussions of attractive policy options for achieving significant emissions reductions from the household sector in a 5-10-year time scale.

Loren Lutzenhiser, Portland State University, Session Moderator

Presentations:
The national potential for emissions reduction from household action—Thomas Dietz, Michigan State University
Achieving the potential for residential energy efficiency—Karen Ehrhardt-Martinez, American Council for an Energy-Efficient Economy
Inducing action through social norms—Wesley Schultz, California State University, San Marcos
Interventions in the supply chain for consumer products and services —Charles Wilson, London School of Economics

Comment and discussion topics
—Economic perspectives on household actions
—Policy opportunities and barriers

Discussion
Adjourn

Session #3—December 4, 2009
Public Acceptance of Energy Technologies

Many current proposals for limiting climate change depend on the development and expeditious deployment of new low-carbon energy supply technologies and new technologies for energy efficiency. Past and recent experiences make clear that public acceptance often slows these processes, sometimes significantly.

This session will present summaries of knowledge about the conditions under which public acceptance issues have and have not significantly slowed implementation of new technologies, particularly energy technologies, and about the effects of different ways of addressing public concerns. Discussion will focus on the implications for the development and deployment of such technologies as wind power, bioenergy technologies, and carbon capture and sequestration. It will surface ideas about how to reconcile pressures for rapid deployment and for well-informed democratic decision making.

Roger Kasperson, Clark University, Session Moderator

Presentations:
Lessons from the past: Governance of emerging energy technologies —Nicholas Pidgeon, Cardiff University
Lessons from the past: Addressing facility siting controversies—Seth Tuler, Social and Environmental Research Institute
Public acceptance issues with renewable energy: offshore wind power —Jeremy Firestone, University of Delaware
Public acceptance issues with carbon capture and storage—Wändi Bruine de Bruin, Carnegie Mellon University

Comment and discussion topics:
—Implications for managing technology development and introduction
—Implications for reaching carbon reduction goals
—Acceptance issues with other new technologies: bioenergy, geoengineering, etc.
—Can government learn the lessons of past energy technologies?

Discussants:
Robert Marlay, U.S. Climate Change Technology Program
Baruch Fischhoff, Carnegie Mellon University

Session #4—December 4, 2009
Organizational Change and the Greening of Business

Businesses are major contributors to climate change through their direct use of energy and land and through their effects on the life cycles of goods and services they use, process, and sell. Behavioral evidence shows that significant resistances exist in business organizations to making transitions to "greener" operations that would be economically rational.

This session will begin with presentations on barriers to change in business that have been identified in organizational theory and research and will then move to a discussion of practical knowledge about the greening of business and about barriers to change. It will end with discussions of what businesses, business organizations, and governments can do to facilitate transitions to greener business practices.

Andrew Hoffman, University of Michigan, Session Moderator

Presentations:
Psychological barriers to organizational change—Max Bazerman, Harvard University
Organizational and institutional barriers to change—Royston Greenwood, University of Alberta
Survey results on barriers to change in businesses—Clay Nesler, Johnson Controls, Inc.

Roundtable discussion among practitioners:
Andre de Fontaine, Markets and Business Strategy Fellow, Pew Center on Global Climate Change
Melissa Lavinson, Pacific Gas and Electric
Clay Nesler, Vice President, Global Energy and Sustainability, Johnson Controls, Inc.

Comment:
Policy possibilities for facilitating organizational change—John Dernbach, Widener University College of Law

List of Participants

David Allen, U.S. Global Change Research Program
Rep. Brian Baird, U.S. Congress, Washington State
Max Bazerman, Harvard University
Bill Blakemore, ABC News
Jay Braitsch, U.S. Department of Education
Wändi Bruine de Bruin, Carnegie Mellon University
Robert Corell, Global Environmental and Technology Foundation
Andre de Fontaine, Pew Center on Global Climate Change
Linda DePugh, The National Academies
John Dernbach, Widener University College of Law
Riley Dunlap, Oklahoma State University
Karen Ehrhardt-Martinez, American Council for an Energy-Efficient Economy
Jeremy Firestone, University of Delaware

Baruch Fischhoff, Carnegie Mellon University
Ilya Fischhoff, U.S. Agency for International Development
Sherrie Forrest, The National Academies
Robert Fri, Resources for the Future, Washington, DC
Jason Gallo, Science and Technology Policy Institute
Elisabeth Graffy, U.S. Department of Interior
Royston Greenwood, University of Alberta
Rachelle Hollander, National Academy of Engineering
Douglas Kaempf, U.S. Department of Energy
Prajwal Kulkarni, U.S. Environmental Protection Agency
Katrina Lassiter, Office of Rep. Brian Baird, U.S. Congress, Washington State
Melissa Lavinson, Pacific Gas and Electric
Linda Lawson, U.S. Department of Transportation
Meredith Blaydes Lilley, University of Delaware
Ed Maibach, George Mason University
Robert Marlay, U.S. Climate Change Technology Program
Tanya Maslak, U.S. Global Research Program
Meg McDonald, Global Issues, Alcoa
Michael Meirovitz, Lewis-Burke Associates, LLC
Claudia Mengelt, The National Academies
Clay Nesler, Johnson Controls, Inc.
Frank Niepold, National Oceanic and Atmospheric Administration
Robert O'Connor, National Science Foundation
Eleonore Pauwels, Woodrow Wilson International Center for Scholars
Nicholas Pidgeon, Cardiff University
Daniel Read, Yale University
David Rejeski, Woodrow Wilson International Center for Scholars
Marcy Rockman, U.S. Environmental Protection Agency
Joseph Ryan, William and Flora Hewlett Foundation
Sarah J. Ryker, Science and Technology Policy Institute
Wesley Schultz, California State University
Stephanie Shipp, Science & Technology Policy Institute
Rachael Shwom, Rutgers University
Paul Stern, The National Academies
Rita Teutonico, National Science Foundation
Seth Tuler, Social and Environmental Research Institute
Louie Tupas, U.S. Department of Agriculture
Kenneth Verosub, U.S. Agency for International Development
Bud Ward, Yale Forum on Climate Change and the Media
Elke Weber, Columbia University
Thomas Webler, Social and Environmental Research Institute
Charles Wilson, London School of Economics

Appendix B

April 2010 Workshop Agenda and List of Participants

Workshop on Adapting to Climate Change:
Insights from the Social Sciences
Agenda and List of Participants
April 8-9, 2010

Thursday, April 8, 2010

Opening remarks by panel chair—Roger Kasperson, Clark University

Overview of the State of the Field
- Addressing strategic and integration challenges of climate change adaptation— Ian Burton, Meteorological Service of Canada and University of Toronto
- Addressing barriers and social challenges of climate change adaptation— Neil Adger, University of East Anglia
- Federal climate change adaptation planning—Maria Blair, White House Council on Environmental Quality
- Adaptation in the America's climate choices study—Claudia Mengelt, Panel on Adapting to the Impacts of Climate Change

Panel Discussion 1: Place-Based Adaptation Cases
- Urban climate adaptation planning: Lessons from the global South—JoAnn Carmin, Massachusetts Institute of Technology
- Climate adaptation: From stories to tools—Amy Luers, Google

Lessons from the RISA experience—Caitlin Simpson and Claudia Nierenberg, National Oceanic and Atmospheric Administration

Panel Discussion 2: Adaptation and Natural Resource Management

Adapting to climate: Learning from the Carolinas water resources sector— Kirstin Dow, University of South Carolina

Knowledge, networks, and water resources— Helen Ingram, University of California, Irvine

Adaptation and marine fisheries management: The Atlantic surfclam case an exemplary or cautionary tale— Bonnie McCay, Rutgers University

Access, articulation, and adaptation to climate change—Ashwini Chhatre, University of Illinois at Urbana-Champaign

Adjourn

Friday, April 9, 2010

Introduction to the day's agenda—Roger Kasperson, Clark University

Panel Discussion 3: Cross-Cutting Issues in Adaptation I

Lessons learned from public health on the process of adaptation—Kristie Ebi, Intergovernmental Panel on Climate Change

The network structure of climate change adaptation: Viewing networks as both opportunities and barriers to successful learning—Adam Henry, West Virginia University

Thoughts on the role of urban areas in adaptation and effective stakeholder-researcher adaptation process—Cynthia Rosenzweig, NASA Goddard Institute for Space Studies

Obstacles and opportunities: Lessons from case studies of adaptation to a changing climate—Neil Leary, Dickinson College

Panel Discussion 4: Cross-Cutting Issues in Adaptation II

Adaptation to climate change through long-term contracts—Howard Kunreuther, Wharton School, University of Pennsylvania

Opportunities and constraints to characterizing and assessing adaptive capacity—Maria Carmen Lemos, University of Michigan

Identifying and overcoming barriers to adaptation: Insights from the trenches of muddling through—Susanne Moser, Susanne Moser Research and Consulting

Comments on Key Questions for the Workshop
Initiating adaptation efforts—Richard Andrews, University of North Carolina, Chapel Hill
Coordinating adaptation efforts—Stewart Cohen, Environment Canada and University of British Columbia
Informing adaptation efforts—Michele Betsill, Colorado State University
Science needs for adaptation efforts—Maria Carmen Lemos, University of Michigan
Managing adaptation efforts—Susanne Moser, Susanne Moser Research and Consulting

Discussion of major insights and issues
Opening comments—Thomas Dietz, Michigan State University

Discussion of next steps
Opening comments—Roger Kasperson, Clark University

Adjourn

List of Participants

Sarah Abdelrahim, National Oceanic and Atmospheric Administration
Neil Adger, University of East Anglia
David Allen, U.S. Global Change Research Program
Richard Andrews, University of North Carolina, Chapel Hill
Peter Banks, National Academy of Sciences
Jainey Bavishi, National Oceanic and Atmospheric Administration
Nancy Beller-Simms, National Oceanic and Atmospheric Administration
Michele Betsill, Colorado State University
Maria Blair, White House Council on Environmental Quality
Ian Burton, Meteorological Service of Canada and University of Toronto
JoAnn Carmin, Massachusetts Institute of Technology
Sarah Carter, Office of Science and Technology Policy
Ashwini Chhatre, University of Illinois at Urbana-Champaign
Stewart Cohen, Environment Canada and University of British Columbia
Linda DePugh, The National Academies
Thomas Dietz, Michigan State University
Kirstin Dow, University of South Carolina
Kristie Ebi, Intergovernmental Panel on Climate Change
Christopher Farley, U.S. Forest Service
Adam Henry, West Virginia University
Helen Ingram, University of California, Irvine

Kathy Jacobs, Office of Science and Technology Policy
Alexa Jay, U.S. Government Accountability Office
Christine Jessup, National Oceanic and Atmospheric Administration
Roger Kasperson, Clark University
Jamie Kruse, National Oceanic and Atmospheric Administration
Howard Kunreuther, Wharton School, University of Pennsylvania
Hadas Kushnir, National Academy of Sciences
Fabien Laurier, U.S. Global Change Research Program
Neil Leary, Dickinson College
Maria Carmen Lemos, University of Michigan
Amy Luers, Google
Tanya Maslak, U.S. Global Change Research Program
Margaret McCalla, National Oceanic and Atmospheric Administration
Bonnie McCay, Rutgers University
Susanne Moser, Susanne Moser Research and Consulting
Claudia Nierenberg, National Oceanic and Atmospheric Administration
Carolyn Olson, U.S. Department of Agriculture Natural Resources Conservation Service
Adam Parris, National Oceanic and Atmospheric Administration
Laura Petes, National Oceanic and Atmospheric Administration
Rick Piltz, Climate Science Watch
Chet Ropelewski, National Oceanic and Atmospheric Administration
Cynthia Rosenzweig, NASA Goddard Institute for Space Studies
Roberto Sanchez-Rodriguez, University of California, Riverside
Michael Savonis, U.S. Department of Transportation
Caitlin Simpson, National Oceanic and Atmospheric Administration
Pamela Stephens, National Science Foundation
Paul Stern, The National Academies
Miron Straf, The National Academies
Rita Teutonico, National Science Foundation
Bob Vallario, U.S. Department of Energy
Robert Verchick, U.S. Environmental Protection Agency
Victoria Wittig, The National Academies

Appendix C

Biographical Sketches of Panel Members and Staff

ROGER E. KASPERSON (*Chair*), is research professor and distinguished scientist at the George Perkins Marsh Institute at Clark University. He has taught at Clark University, the University of Connecticut, and Michigan State University. His expertise is in risk analysis, global environmental change, and environmental policy. Dr. Kasperson is a Fellow of the American Association for the Advancement of Science (AAAS) and the Society for Risk Analysis. He has served on numerous committees of the National Research Council (NRC). He chaired the International Geographical Commission on Critical Situations/Regions in Global Environmental Change and has served on the Science Advisory Board of the U.S. Environmental Protection Agency (EPA). He is cochair of the Scientific Advisory Committee of the Potsdam Institute for Climate Change, and is on the Executive Steering Committee of the START Programme of the International Geosphere Biosphere Program. He is a member of the National Academy of Sciences and the American Academy of Arts and Sciences. He has authored or coedited 22 books and monographs and more than 143 articles or chapters in scholarly journals or books and has served on numerous editorial boards for scholarly journals. From 2000 to 2004, Kasperson was executive director of the Stockholm Environment Institute in Sweden. He was a coordinating lead author of the vulnerability and synthesis chapters of the Conditions and Trends volume of the Millennium Ecosystems Assessment and a member of the core writing team for the Synthesis of the overall assessment. Kasperson has been honored by the Association of American Geographers for his hazards research and in 2006 he was the recipient of the Distinguished Achievement Award of the Society for Risk Analysis. In

2007, he was appointed as associate scientist at the National Center for Atmospheric Research (NCAR) in the United States. He received his Ph.D. from the University of Chicago.

RICHARD N. ANDREWS is professor of environmental policy in the Department of Public Policy, the Department of City and Regional Planning, and the Curriculum for the Environment and Ecology in the College of Arts and Sciences, and in the Department of Environmental Sciences and Engineering in the Gillings School of Global Public Health at the University of North Carolina (UNC), Chapel Hill. His research and teaching are on environmental policy in the United States and worldwide, including books on the history of U.S. environmental policy and on the National Environmental Policy Act, and research grants on environmental policy innovations in the United States, the Czech Republic, and Thailand. Beyond the university, he has twice chaired the Section on Societal Impacts of Sciences and Engineering of AAAS, and also has served as a member of its Committee on Science, Engineering, and Public Policy. He has chaired or served on study committees for the NRC, the Science Advisory Board of EPA, the National Academy of Public Administration, and the Congressional Office of Technology Assessment. He was principal environmental staff member for the 1984 *The Future of North Carolina* study, which was commissioned by the governor. A member of the UNC faculty since 1981, Andrews served as chair of the UNC faculty from 1997 to 2000. Before joining the Carolina faculty, he taught for 9 years in the University of Michigan's School of Natural Resources, and served as a Peace Corps volunteer and an analyst for the U.S. Office of Management and Budget. He earned the AB degree from Yale, and the Ph.D. and a professional master's degree from UNC's Department of City and Regional Planning.

MICHELE M. BETSILL is associate professor of political science at Colorado State University. Previously, she was a postdoctoral fellow with the Global Environmental Assessment project at Harvard's John F. Kennedy School of Government, the Colorado State University faculty in residence at the Central and East European Studies Program at the Economics University of Prague, and a visiting scientist at NCAR. Her research focuses on global environmental governance, particularly related to the issue of climate change and more specifically the multilevel nature of climate change governance, including levels of political jurisdiction from the local to the global and across the public and private sectors. Her current projects investigate the ways that institutions and actors interact across various tiers and spheres of governance and the implications for addressing the threat of climate change and for understanding of global environmental governance. She is coauthor of *Cities and Climate Change: Urban Sustainability and Global Environmental*

Governance (Routledge, 2003) and coeditor of *NGO Diplomacy: The Influence of Nongovernmental Organizations in International Environmental Negotiations* (MIT Press, 2008) and numerous peer-reviewed articles. She received her B.A. from DePauw University, M.A. degrees from the University of Denver and the University of Colorado, Boulder, and a Ph.D. in political science from the University of Colorado, Boulder.

STEWART J. COHEN is senior researcher with the Adaptation and Impacts Research Section of Environment Canada, and an adjunct professor with the Department of Forest Resources Management of the University of British Columbia. Dr. Cohen's research interests are in climate change impacts and adaptation at the regional scale, and exploring how climate change can affect sustainable development. Recent and ongoing studies include climate change and water management in the Okanagan region of British Columbia, climate change visualization, and methods for incorporating climate change adaptation into municipal planning and forest management. He is currently a member of the advisory committee for the Columbia Basin Trust program, Communities Adapting to Climate Change. Previously, he led the Mackenzie Basin Impact Study, a 7-year effort focused on climate change impacts in the western Canadian Arctic, completed in 1997. His earlier work included research on impacts in the Great Lakes and Saskatchewan River Basins, and advising the Canadian Climate Impacts and Adaptation Research Network. He has been a lead author for the Intergovernmental Panel on Climate Change Third and Fourth Assessment Reports, and the U.S. Climate Change Science Program report, *Global Climate Change Impacts in the United States*, published in 2009. He also published a textbook (with Melissa Waddell), entitled *Climate Change in the 21st Century*, a study guide for promoting interdisciplinary collaboration. Dr. Cohen is a geographer having received his B.Sc., M.Sc., and Ph.D. from McGill University, University of Alberta, and University of Illinois, respectively.

THOMAS DIETZ is professor of sociology and of crop and soil sciences, director of the Environmental Science and Policy Program, and assistant vice president for Environmental Research at Michigan State University. He is a fellow of AAAS, a Danforth Fellow, past-president of the Society for Human Ecology and has received the Distinguished Contribution Award from the Section on Environment, Technology and Society of the American Sociological Association and the Sustainability Science Award of the Ecological Society of America. His research interests include the role of deliberation in environmental decision making, the human dimensions of global environmental change and cultural evolution. He holds a B.G.S. from Kent State University and a Ph.D. in Ecology from the University of California, Davis.

ANDREW J. HOFFMAN is the Holcim (U.S.) professor of sustainable enterprise; associate professor of management and organizations; associate professor of natural resources; and associate director of the Erb Institute for Global Sustainable Enterprise, at the University of Michigan. He studies organizational culture, values, and behavior, with a particular emphasis on corporate strategies for addressing climate change. Previously, he was associate professor of organizational behavior at the Boston University School of Management; was a senior fellow at the Meridian Institute working on promoting discussion among senior industry, government and nongovernmental representatives; and developing a training program for senior chemical industry executives on constructive engagement with external stakeholders. He also served previously as an analyst for the Amoco Oil Company, modeling the expected costs and potential strategies for dealing with the Clean Air Act Amendments and other environmental statutes. Dr. Hoffman has written numerous books and articles about corporate strategies for addressing climate change, and has organized and moderated conferences on Corporate Strategies That Address Climate Change; Reframing the Climate Change Debate; and Senior Level Dialogues on Climate Change Policy; bringing together senior executives from business, government and the environmental community to discuss the scientific, strategic and policy implications of controls on greenhouse gas emissions. He has a Ph.D. (interdepartmental degree) from MIT from the Alfred P. Sloan School of Management and the Department of Civil and Environmental Engineering.

ANTHONY LEISEROWITZ is director of the Yale Project on Climate Change and a research scientist at the School of Forestry and Environmental Studies at Yale University. He is also a principal investigator at the Center for Research on Environmental Decisions at Columbia University. He is an expert on American and international public opinion on global warming, including public perception of climate change risks, support and opposition for climate policies, and willingness to make individual behavioral change. His research investigates the psychological, cultural, political, and geographic factors that drive public environmental perception and behavior. He has conducted survey, experimental, and field research at scales ranging from the global to the local, including international studies, the United States, individual states (Alaska and Florida), municipalities (New York City), and with the Inupiaq Eskimo of Northwest Alaska. He also recently conducted the first empirical assessment of worldwide public values, attitudes, and behaviors regarding global sustainability, including environmental protection, economic growth, and human development. He has served as a consultant to Harvard's John F. Kennedy School of Government, the United Nations Development Program, the Gallup World Poll, the Global Roundtable on Climate Change at the Earth Institute (Columbia University), and the World Economic Forum.

LOREN LUTZENHISER is professor of urban studies and planning at Portland State University. Dr. Lutzenhiser's teaching interests include environmental policy and practice, energy behavior and climate, technological change, urban environmental sustainability, and social research methods. His research focuses on the environmental impacts of socio-technical systems, particularly how urban energy/resource use is linked to global environmental change. Particular studies have considered variations across households in energy consumption practices, how energy-using goods are procured by government agencies, how commercial real estate markets work to develop both poorly-performing and environmentally exceptional buildings, and how the "greening" of business, may be influenced by local sustainability movements and business actors. He recently completed a major study for the California Energy Commission reporting on the behavior of households, businesses and governments in the aftermath of that state's 2001 electricity deregulation crisis. He is currently exploring the relationships between household natural gas, electricity, gasoline, and water usage. He holds a Ph.D. in sociology.

SUSANNE C. MOSER is director and principal researcher of Susanne Moser Research and Consulting. Previously, she was a scientist at the Institute for Study of Science and Environment at NCAR in Boulder, Colorado. She has also served as staff scientist at the Union of Concerned Scientists, a visiting assistant professor at Clark University, and a fellow in the Global Environmental Assessment Project at Harvard University. Her research interests include the impacts of global environmental change, especially in the coastal, public health, and forest sectors; societal responses to environmental hazards in the face of uncertainty; the use of science to support policy and decision making; and the effective communication of climate change to facilitate social change. Her current work focuses on developing adaptation strategies to climate change at local and state levels, identifying ways to promote community resilience, and building decision support systems. She is a fellow of the Aldo Leopold and Donella Meadows Leadership Programs. She received a diploma in applied physical geography from the University of Trier in Germany and M.A. and Ph.D. degrees in geography from Clark University.

PAUL C. STERN is a principal staff officer at the NRC/National Academy of Sciences, director of its Committee on the Human Dimensions of Global Change, and study director for this panel. His research interests include the determinants of environmentally significant behavior, particularly at the individual level; participatory processes for informing environmental decision making; processes for informing environmental decisions; and the governance of environmental resources and risks. He is coauthor of the textbook *Environmental Problems and Human Behavior* (2nd ed., 2002); coeditor of numerous NRC publications, including *Public Participation in Environmental Assessment and Decision Making* (2008), *Decision Making for the Environment: Social and Behavioral Science Priorities* (2005), *The Drama of the Commons* (2002), *Making Climate Forecasts Matter* (1999), *Environmentally Significant Consumption: Research Directions* (1997), *Understanding Risk* (1996), *Global Environmental Change: Understanding the Human Dimensions* (1992), and *Energy Use: The Human Dimension* (1984). He directed the study that produced *Informing Decisions in a Changing Climate* (2009). He coauthored the article "The Struggle to Govern the Commons," which was published in *Science* in 2003 and won the 2005 Sustainability Science Award from the Ecological Society of America. He is a fellow of AAAS and the American Psychological Association. He holds a B.A. from Amherst College and an M.A. and Ph.D. from Clark University, all in psychology.

GARY W. YOHE is John E. Andrus professor of economics and director of the John E. Andrus Public Affairs Center at Wesleyan University. His research focuses on adaptation and the potential damage from global climate change. It examines micro-responses to investigate the degree to which assuming efficient markets biases the estimates of cost and/or limits the range of potential adaptation; estimations of reduced-form cost functions when data are scarce; and the role of uncertainty and the search for robust and/or hedging strategies in formulating policy. He holds a Ph.D. in economics from Yale University.

'I struggled to put it down . . . I hope everyone discovers Lindsey Barraclough because her writing is great escapism' *TES*

'*Long Lankin* is truly spine-chilling. A real sense of menace pervades the book and the cold dampness of the abandoned church and Ida's neglected home feel almost tangible . . . I loved this book. It's such an impressive debut. Every element is spot on – from the elegant prose, through the realistic portrayal of various aspects of family life, the three-dimensional characters and the occasional comic set-piece, to the supernatural horror underpinning it, which is absolutely chilling. Highly recommended' *The Bookbag*

'This is an absolutely stunning debut novel . . . The tension is almost unbearable and the atmosphere and setting genuinely spine-tingling. Be warned, once you pick it up you will not want to be disturbed or be able to put it down' *Reading Zone*

'A uniquely creepy read and one of the most original YA novels I have ever read. Totally unputdownable' *www.overflowinglibrary.com*

www.totallyrandombooks.c

'Not for the faint-hearted, this mesmerising tale generates goosebumps on almost every page'
Booktrust

'Not only was *Long Lankin* one of those books I wanted to reread as soon as I'd finished it, but I also wanted to thrust a copy into the hands of everyone I know and demand they read it. The story is as creepy and atmospheric as the cover suggests and I completely and utterly fell in love with it. I honestly don't know where to begin. I found the story completely mesmerising . . . I can't fault this one at all. A stunning debut'
I Want to Read That

'Both the song and the novel tell a haunting story that can definitely make the hairs rise on your skin . . . As much a delight to read as it is scary. There are elements of family drama, of history and of the supernatural, and I really got the sense of a typical English village somewhere in a marshy area where superstitions and legends are often more than just a rumour'
The Bookette

'Inescapably haunting and beautiful . . . Truly original' *Portrait of a Woman*

'A brilliant debut novel with a beautifully crafted and perfectly executed story. Told through the different characters' voices, the narrative is skilfully balanced, giving you laughs one minute and chills the next. The characters feel as real as the small, rural village Barraclough conjures up' *Sugarscape.com*

'I am utterly obsessed with *Long Lankin* . . . A brilliantly written, psychologically harrowing horror'
Writing From the Tub